H. Newell Martin

The human body

An elementary text-book of anatomy, physiology and hygiene

H. Newell Martin

The human body
An elementary text-book of anatomy, physiology and hygiene

ISBN/EAN: 9783742862822

Manufactured in Europe, USA, Canada, Australia, Japa

Cover: Foto ©berggeist007 / pixelio.de

Manufactured and distributed by brebook publishing software
(www.brebook.com)

H. Newell Martin

The human body

Edition with Special Treatment of Alcohol and other
Narcotics

THE HUMAN BODY

AN

ELEMENTARY TEXT-BOOK OF ANATOMY, PHYSIOLOGY, AND HYGIENE

BY

H. NEWELL MARTIN, D.Sc., M.A., F.R.S.

PROFESSOR OF BIOLOGY IN THE JOHNS HOPKINS UNIVERSITY

NEW YORK
HENRY HOLT AND COMPANY
1892

ENDORSEMENT.

We have examined the valuable book, "THE HUMAN BODY," specially revised as to Alcohol and other Narcotics, and heartily commend it.

Its clear statement of the scientific facts in the case constitutes a powerful argument against the use of alcoholic drinks and other narcotics, which cannot fail to influence students for an intelligent sobriety.

MARY H. HUNT,

National and International Superintendent of the Scientific Department of the Woman's Christian Temperance Union; Life Director of the National Educational Association.

ADVISORY BOARD :

JOSEPH COOK,   WILLIAM E. SHELDON,
ALBERT H. PLUMB, D.D.,   DANIEL DORCHESTER, D.D.

PREFACE.

THE Physiology and Hygiene of this text-book are, in general, identical with that of the "Briefer Course" edition of the "Human Body" published in the "American Science Series." In the present edition the chemical phraseology has been simplified in consequence of the requests of some teachers. A chapter has been added on the subjects of Fermentation and Distillation. The description of the effects of alcoholic drinks and other narcotics on the human system has been amplified, and separated into sections appended to various chapters, instead of being collected in a chapter by itself as in the "Briefer Course" edition. These changes bring the book into harmony with the laws requiring the facts concerning these substances to be taught in public schools.

H. NEWELL MARTIN.

JOHNS HOPKINS UNIVERSITY,
    January 1, 1890.

CONTENTS.

CHAPTER I.

THE GENERAL STRUCTURE AND ARRANGEMENT OF THE HUMAN
BODY.

CHAPTER II.

THE MICROSCOPICAL AND CHEMICAL COMPOSITION OF THE BODY.

CHAPTER III.

PUTREFACTION, FERMENTATION, DISTILLATION, AND THE NATURE
OF ALCOHOL.

CHAPTER IV.

THE SKELETON.

CHAPTER XI.

FOODS.

CHAPTER XII.

THE DIGESTIVE ORGANS.

CHAPTER XIII.

DIGESTION.

CHAPTER XIV.

BLOOD AND LYMPH.

CHAPTER XV.

THE ANATOMY OF THE CIRCULATORY ORGANS.

The organs of circulation—Functions of the main parts of the

THE HUMAN BODY.

CHAPTER I.

THE GENERAL STRUCTURE AND ARRANGEMENT OF THE HUMAN BODY.

Human Physiology is that department of science which has for its object the discovery and accurate description of the properties and actions of the living healthy human body, and the finding of the uses or (as physiologists call them) the *functions*, of its various parts. Physiologists endeavor to discover what the body, as a whole, does while alive, and what each part of it does, and how it does it; also under what conditions the work of the body is best performed; and, in connection with this last aim, to furnish the basis of *Hygiene*, the science which is concerned with the laws and conditions of health.

What is human physiology? What is a function? What do physiologists endeavor to discover? What is meant by Hygiene?

1

Anatomy.—Clearly, the first step to be taken towards finding out the use and mode of working of each part of the body is to find out what the parts are ; this study is known as *Human Anatomy.* Examined merely from the exterior, the body is quite a complicated structure : we can all see for ourselves, head and neck, trunk and limbs, and even many smaller but quite distinct parts entering into the formation of these larger ones, as eyes, nose, ears, and mouth ; arm, forearm, and hand ; thigh, leg, and foot. This knowledge of its complexity, which we may all arrive at by looking on the outside of the body, is vastly extended when it is dissected and its interior examined ; we then learn that it is made up of many hundreds of diverse parts, each having its own structure, and form, and purpose, but all harmoniously working together in health.

Summary.—*Anatomy* is concerned with the form and structure and connections of the parts of the body. *Physiology* with the uses of the parts, and the ways in which they work. *Hygiene* with the conditions of life which promote the health of the body.

Microscopic Anatomy or Histology.—When we examine the body from its exterior, we observe that a number of different materials enter into its formation. Hairs, nails, skin, and teeth are quite different substances ; by feeling through the skin we find harder and softer solid

What is human anatomy ? Give illustrations of the complexity of the body in structure? Is its internal structure as varied as its external?

State in a few words the subject matters of the sciences of Human Anatomy, Physiology, and Hygiene.

Give examples of the variety of substances entering into the composition of the body. What may we feel through the skin ?

masses under it; while the blood which flows from a cut finger, and the saliva which moistens the mouth, show clearly that liquids exist in the body.

If we were to go farther and examine closely, outside and inside, any one part of the body, the hand for instance, we should find it made up of quite a number of different materials. On its exterior we see *skin* and *nails;* if the skin were dissected off we should find under it more or less yellowish-white *fat;* beneath the fat would lie a number of red soft masses, the *muscles* (answering to what we call the *lean* of meat); under the muscles, again, would be hard, rigid, whitish *bones;* at the finger joints, where the ends of different bones lie close together, we should find them covered by still another substance, *gristle* or *cartilage.* Finally, binding skin and fat and muscles and bone together, we should discover a tough stringy material, quite different from all of them, and which, since it unites all the rest, is called *connective tissue.* If we took any other portion of the body we should arrive at a similar result; it, too, would be made up of a number of different materials, which materials might, as in the case of the foot, be identical with those found in the hand but arranged together in a different way so as to perform another function (just as wood and nails may be used to build a house or a bridge, but are put together in a different manner in the two cases); or we might, as in the eye, find in addition to some materials found in the hand others quite unlike any of them.

What different materials do we see on looking at a hand? What others would be found on dissecting away the skin? What is the technical name for the lean of meat? What is the technical name of gristle? What is connective tissue? Are other parts of the body besides the hand made up of different substances? Illustrate by an example?

The branch of anatomy which deals with the characters of the materials used in the construction of the parts of the body is called *histology*, or, since it is mainly carried on with the aid of the microscope, *microscopic anatomy*.

Tissues.—Each of the different primary building materials which can be distinguished, either with or without the microscope, as entering into the construction of the body, is called a *tissue ;* we speak, for example, of muscular tissue, fatty tissue, bony tissue, cartilaginous tissue, and so forth ; each tissue has certain properties in which it differs from all the rest, and which it preserves in whatever part of the body it may be found. It also is characterized by certain appearances when examined with a microscope, which are the same for the same tissue no matter where it is found. The total number of important tissues is not great ; the variety in structure and use which we find in the parts of the body depends mainly on the diverse ways in which the same tissues are combined together, over and over again, in different parts.

Organs.—A portion of the body composed of several tissues, and specially fitted for the performance of a particular duty or function, is called an *organ ;* thus, the hand is an organ of prehension ; the eye, the organ of sight ; the stomach, an organ of digestion ; and so forth.

Summary.—The human body is made up of a limited number of *tissues ;** each tissue has a characteristic ap-

What is histology?   Give another name for it.

What is a tissue?   Give examples.   How do tissues differ?   Are there a large number of important tissues ?   How is the variety in structure of the parts of the body produced?

What is an organ?   Give examples.   Of what is the body made up ?

---

* The various tissues of the body will be considered in more detail subsequently: the more important are—1. Bony tissue. 2. Cartilaginous tissue. 3. White fibrous

pearance, by which it can be recognized with the microscope, and some one or more distinctive properties which fit it for some special use ; thus, it may be very tough, and suited for binding other parts together ; or rigid, and adapted to preserve the shape of the body ; or have the power of changing its length and be useful for moving parts to which its ends are attached.

The tissues are variously combined to form the *organs* of the body, of which there are very many, differing in size, shape, and structure ; some organs contain only a few tissues ; others, a great many ; some possess only tissues which are found also in other organs, others contain one or more tissues peculiar to themselves ; but wherever an organ is found, it is constructed and placed with reference to the performance of some duty ; the organs are the machines which are found in the factory represented by the body, and the tissues are the materials used in building the machines ; or, using another illustration, we may, with Longfellow, compare the body to a dwelling-house ; and then go on to liken the tissues to the brick, stone, mortar, wood, iron, and glass, used in building it ; and the organs to the walls, floors, ceilings, doors, and windows, which, made by combining the primary building materials in different ways, have each a purpose of their own, and all together make the house.

How are tissues recognized ? Give examples of differences in properties of various tissues.

How do organs differ from one another ? In what do all organs agree ? Illustrate the relation of organs and tissues to the body as a whole ?

---

connective tissue. 4. Yellow elastic tissue. 5. Glandular tissue, of which there are many varieties. 6. Respiratory tissues. 7. Fatty tissue. 8. Sense-organ or irritable tissues. 9. Nerve cell tissue. 10. Nerve fiber tissue. 11. Striped muscular tissue, 12. Unstriped muscular tissue. 13. Epidermic and epithelial tissue.

The general plan on which the Body is constructed.—
When we desire to gain a general idea of the structural
plan of any object we examine, if possible, sections made
through it in different directions ; the botanist cuts the
stem of the plant he is examining lengthwise and cross-
wise, and studies the surfaces thus laid bare ; a geologist,
investigating the structure of any portion of the earth's
crust, endeavors to find exposed surfaces in cañons, in
railway cuttings, and so forth, where he may see the strata
exposed in their natural relative positions ; and the archi-
tect draws plans which show to his clients sections of the
building which he proposes to erect for them ; so, also, the
best method of getting a good general idea of the way in
which the parts of the human body are put together is
to study them as laid bare by cuts made in different directions ; this gives us a general outline and the details may
be filled in afterwards.

If the whole body were divided from the crown of
the head to the lower end of the trunk, and exactly in the
middle line, so as to separate it into right and left halves,
we should see something like Fig. 1, if we looked at the cut
surface of the right half. Such a section shows us, first,
that the body fundamentally consists of two tubes or cavities, separated by a solid bony partition. The larger cavity, *b, c,* known as the *ventral* or *hæmal cavity,* lies on
the front side, and contains the greater part of the organs

How do we start by preference to gain a knowledge of the structural plan of any mass of matter? Give examples. Apply to
the study of human anatomy.

What should we see on examining the cut surface of a human
body divided into right and left halves? What organs lie in the
hæmal cavity? What in the neural? How far does the hæmal
cavity extend toward the head ?

concerned in keeping up the blood flow (organs of circulation), in breathing (organs of respiration) and in digesting food (organs of digestion). It does not reach up into the neck, but is entirely confined to the trunk. The smaller cavity, *a*, *a'*, is tubular in the trunk region, but passes on through the neck, and widens out in the skull; it is known as the *dorsal* or *neural cavity*, and contains the most important nervous organs, the brain, *N'*, and spinal cord, *N*. In the partition between the two cavities is a stout bony column, the backbone or spine, *e*, *e*, which is made up of a number of short thick bones piled one on the top of another.

Man is a vertebrate animal. —The presence of these two chambers with the solid partition between them is a primary fact in the anatomy of the body; it shows that man is a *vertebrate animal*, that is to say, is a back-boned animal, and belongs to the same great

What lies between the hæmal and neural cavities? Of what is the spine composed?

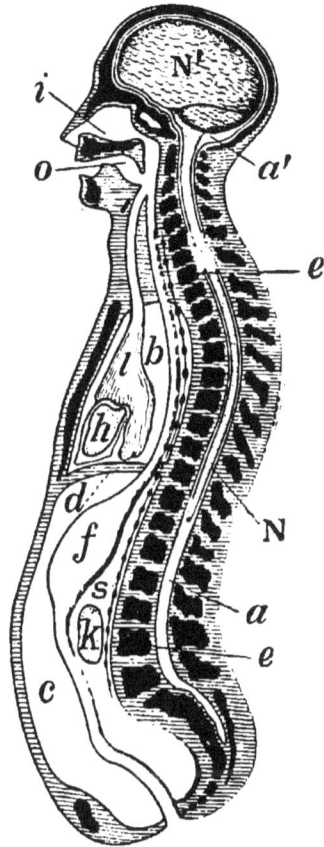

Fig. 1.—Diagrammatic longitudinal section of the body. *a*, the neural tube, with its upper enlargement in the skull cavity at *a'*; *N*, the spinal cord; *N'*, the brain; *ee*. vertebræ forming the solid partition between the dorsal and ventral cavities; *b*, the pleural, and, *c*, the abdominal division of the ventral cavity, separated from one another by the diaphragm, *d*; *i*, the nasal, and *o*, the mouth chamber, opening behind into the pharynx, from which one tube leads to the lungs, *l*, and another to the stomach, *f*; *h*, the heart; *k*, a kidney; *s*, the sympathetic nervous chain. From the stomach, *f*, the intestinal tube leads through the abdominal cavity to the posterior opening of the alimentary canal.

group as fishes, reptiles, birds, and beasts* : sea ane-
mones, clams, and insects are invertebrate animals, and
built on quite different plans ; sections made through
any of them from the head to the opposite end, would
show nothing like those two main cavities with a backbone
between them which exist in our own bodies.

Contents of the two chief cavities of the body.—Examination of Fig. 1 shows that the ventral cavity is entirely
closed itself, though some things which lie in it are hollow
and communicate with the exterior. On the head we
find the nose, *i*, and the mouth, *o*, opening on the *ventral side ;* that is on that surface of the body next which
the hæmal cavity lies. The nose chamber joins the mouth
chamber at the throat, and from the throat two tubes
run down through the neck and enter the ventral cavity.
One of these tubes, placed on the ventral side of the other,
is the *windpipe,* and leads to the lungs, *l ;* the other is the
gullet, and leads to the stomach, *f.* From the stomach,
another tube, *the intestine,* leads to the outside again at
the lower or posterior† end of the trunk. Mouth, throat,

What fact in man's anatomy makes him a vertebrate animal?
Name some other vertebrate animals. Name some invertebrates.
How would sections made through invertebrates differ essentially
from similar sections made through a man?
  Is the ventral cavity open? Do any smaller cavities in it open on
the exterior of the body? What openings do we find on the head?
What is the windpipe? To where does it lead? What is the intestine? What parts constitute the alimentary canal? Does this canal
lie entirely within the hæmal cavity?

---

* The main groups in which animals are arranged are—1. *Vertebrata,* or backboned
animals. 2. *Mollusca,* including snails, slugs, clams, oysters. &c. 3. *Arthropoda,*
including flies, moths, beetles, centipedes, lobsters, spiders, &c. 4. *Vermes,* including worms of various kinds. 5. *Echinodermata* (hedgehog-skinned animals), including sea urchins, star fishes, &c. 6. *Cœlenterata,* the sea-anemones and their
allies. 7. *The Protozoa ;* all microscopic and very simple in structure.

† In anatomy the head end of an animal is spoken of as anterior, and the opposite end as posterior, no matter what may be the natural standing position of the
creature.

gullet, stomach, and intestine, together form *the alimentary canal,* which, as we see, begins on the head quite
above or anterior to the ventral cavity ; then, at the bottom of the neck, enters the ventral cavity and runs on
through it, to pass out again posteriorly ; just as a tube
might pass quite through a box, in at one end and out at
the other, without opening into it at all. In addition to
the lungs and the greater part of the alimentary canal,
the ventral cavity contains several other things of which
we shall have more to say presently ; among the more
important of them are, *the heart, h ; the kidneys, k ;* the
sympathetic nerve centers, s ; and several large organs
making juices which are conveyed by tubes into the alimentary canal and assist in digesting our food.

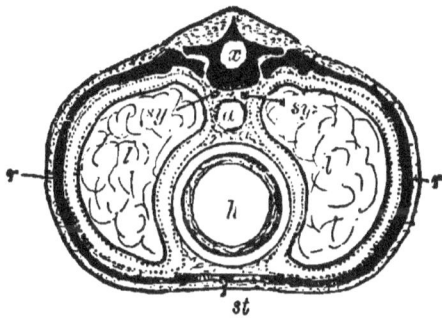

FIG. 2.—A diagrammatic section across the body in the chest region. *x,* the dorsal
tube, which contains the spinal cord ; the black mass surrounding it is a vertebra ;
a, the gullet, a part of the alimentary canal ; *h,* the heart ; *sy,* sympathetic nervous
system ; *ll,* lungs ; the dotted lines around them are the pleuræ ; *rr,* ribs ; *st,* the
breastbone.

If we examined a section made across the trunk of the
body, say about the level of the middle of the chest (Fig.
2), we would find, on the dorsal side, the neural tube, *x,*
cut across, and in it the spinal cord, which is not repre-

Name some organs, in addition to parts of the alimentary canal,
which are found in the ventral cavity. Describe what would be
seen on a section made across the body about the middle of the
chest.

sented in the figure. The thick black mass below the neural tube is part of the spinal column ; bounded by this dorsally, by a rib, *r, r*, on each side, and by the breastbone, *st*, on the ventral side (below in the figure) is the hæmal cavity, containing the lungs, *l, l;* the heart, *h;* the gullet, *a;* and the sympathetic centers, *sy*.

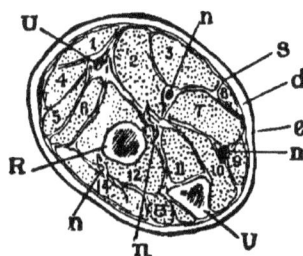

FIG. 3.—A section across the forearm a short distance below the elbow-joint. *R* and *U*, its two supporting bones, the radius and ulna; *e*, the epidermis and *d*, the dermis, of the skin; the latter is continuous below with bands of connective tissue, *s*, which penetrate between and invest the muscles, which are indicated by numbers, *n, n*, nerves and vessels.

The Limbs.—If, instead of the trunk of the body, our section were made across one of the limbs, we should find no such arrangement of cavities on each side of a bony axis. The limbs have a supporting axis made of one or more bones (as seen at *U* and *R*, Fig. 3, which represents a section made across the forearm near the elbow joint), but around this axis soft parts, chiefly muscles, are closely packed ; and the whole, like the trunk, is enveloped by skin. The only cavities in the limbs are branching tubes, which are filled during life either with *blood*, or a watery-looking liquid known as *lymph*. These tubes, the *blood* and *lymph vessels* respectively, are not, however, characteristic of the limbs, for they also exist in abundance in head, neck and trunk.

How do the limbs differ from the trunk? How is each limb supported? Describe the parts exposed on a cross section of the forearm? What cavities exist in a limb? What do they contain? Are they found in other parts of the body?

Man's place among Vertebrates.—It must be clear to every one that although man's structural plan in its broad features, simply indicates that he is a vertebrate animal, yet he is much more like some vertebrates than others. The hair covering more or less of his body, and the organs which produce milk for the nourishment of the infant by its mother, are absent entirely in fishes, reptiles, and birds, but are possessed by ordinary four-footed beasts and by whales, bats, and monkeys. The organs which form milk are the *mammary glands,* and all kinds of animals whose females possess them are known as *Mammalia**: man is, therefore, a Mammal. In internal structure one of the most important characters of the Mammalia is the presence of a cross-partition, called the *midriff* or *diaphragm,* which separates the hæmal cavity into an anterior and a posterior division. This partition is shown at *d* in Fig. 1, where it is seen to divide the ventral cavity into an upper and a lower story; the upper or anterior is the *chest* or *thorax cavity;* the lower or posterior, the *abdominal cavity.* The chest contains the heart, lungs, and most of the gullet; the abdomen contains the lower

In what external characters does the human body differ from that of fishes, reptiles, and birds? Name some animals which agree with mankind in the possession of these characters? What are the mammary glands? What is meant by Mammalia? To what division of vertebrate animals does man belong?

Point out a fact in internal structure in which the Mammalia differ from other vertebrates? Where does the diaphragm lie? What is the name of the cavity above it? What is the name of the cavity below it? Name some organs lying in the thorax. Some placed in the abdomen. Some which run through both.

---

* Zoologists classify vertebrate animals in five groups. 1. *Pisces,* including all true fishes, as sharks, eels, salmons, shad, perch, &c., but excluding the so-called shellfish, as oysters, clams, and lobsters, which are not vertebrates at all. 2. *Amphibia,* frogs, toads, newts, salamanders &c. 3. *Reptilia,* lizards, alligators, turtles, snakes. 4. *Aves,* birds. 5. *Mammalia.*

end of the gullet (which pierces the diaphragm), the stomach, the intestine, the kidneys, and most of the organs making digestive liquids. The sympathetic nerve

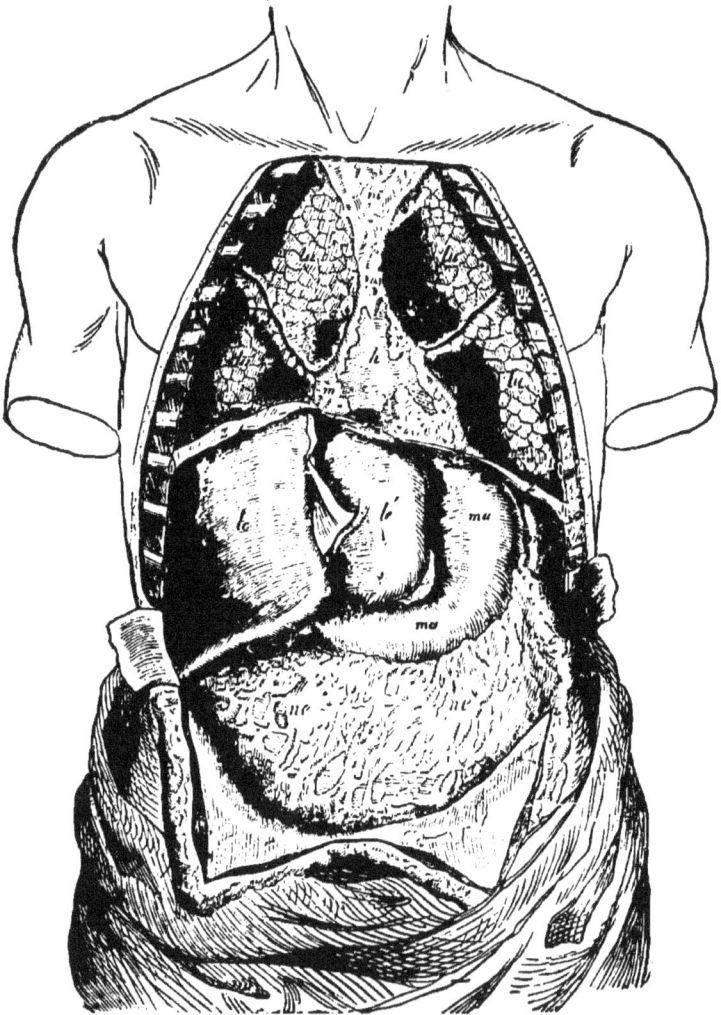

Fig. 4.—The body opened from the front to show the contents of its ventral cavity. *lu*, lungs; *h*, heart, partly covered by other things; *le, le'*, right and left liver lobes respectively; *ma*, stomach; *ne*, the great omentum, a membrane containing fat which hangs down from the posterior border of the stomach and covers the intestines; *nu*, spleen; *zz*, diaphragm.

centers run through both abdomen and chest, and extend beyond the latter into the neck.

The ventral cavity, opened from the front, but with its contents undisturbed, is shown in figure 4. We there see the edge of the diaphragm, *z, z* ; above this, in the chest, the lungs, *lu, lu,* and the heart, *h* ; the latter partly covered by other things. Below the diaphragm is the abdominal cavity, containing in its upper part the liver, *le, le'*; the stomach, *ma* ; and the spleen, *mi* ; hanging down like an apron from the lower border of the stomach is the *omentum, ne, ne,* which lies over and conceals the intestines.

Summary.—Man is a vertebrate animal, because his body presents dorsal and ventral cavities separated from one another by a hard partition ; the dorsal cavity contains the brain and spinal cord, and reaches into the head; the ventral cavity stops at the bottom of the neck and contains the main organs of circulation, respiration, and digestion.

Man belongs to that subdivision of vertebrates known as Mammalia (1) because more or less of his surface is covered by hair ; (2) because of the presence of mammary glands ; (3) because the ventral cavity is completely separated by the diaphragm into thorax and abdomen.

That man is intellectually incomparably superior to any other animal, and stands supreme in the world, can

Name the parts seen when the front wall of a man's trunk is cut away. Describe the relative positions of these parts. On what anatomical grounds do we call man a vertebrate animal? What lies in the dorsal or neural cavity? How far does the upper end of this cavity reach? What organs lie in the ventral cavity? Where does its upper limit lie? Why is man a Mammal? In what is man superior to any other animal? From what point of view have anatomists to regard man's body? What sort of facts do they take into account in assigning man's position among animals?

be doubted by no one; still greater is his supremacy when
we consider his power of forming conceptions of right and
wrong, and his knowledge of moral responsibility. But
anatomists have only to deal with man's body as a material object, and as such they classify it among other animal bodies according to the greater or less resemblances
or differences which are found between it and them.*

---

* It will be found very useful to accompany the teaching of this chapter with
a demonstration on the body of a dead rat, kitten, or puppy. On opening the
body the chest and abdominal cavities will be readily shown, and also the main
organs in them. Then, on opening the skull, the brain will be seen, and on
cutting across the spinal column with strong scissors, the slender soft spinal
cord lying in its tube will come into view.

CHAPTER II.

THE MICROSCOPICAL AND CHEMICAL COMPOSITION OF THE BODY.

What the tissues are like.—Having gained some idea of how the larger parts of the body are arranged we may next inquire what the tissues, its smallest parts which are combined to make the larger, are like. The simplest tissues are known as *cells;* * they are so small that a separated cell can only be seen with the help of a microscope. In a fully formed cell (Fig. 5) we find three parts : (1) a *cell body* made up of a soft granular substance ; (2) a smaller and less granular *cell nucleus* imbedded in the cell body ; and (3) a tiny dot, the *nucleolus*, lying in the nucleus. Cells vary much in form and size, though all are very small. A good many, like those represented at *b*, float in our blood, and are more or less rounded. In other places cells are flattened to form thin scales as those in Fig. 6, which represents cells scraped

FIG. 5.—Forms of cells from the body.

Of what are the larger parts of the body made up? What are the simplest tissues? What instrument must we employ in order to see them? Describe the structure of a cell. Describe some different forms of cells.

---

* So called from an old belief that they were little bags or chambers. Most cells are really solid or semi-solid throughout.

from the inside of the wall of the abdomen. Elsewhere
we find cells elongated, as *c*, Fig. 5 ; if this goes on to any
great extent we get a long slender thread which is called
a *fiber ;* but very often fibers are made by a number of
cells, all elongating a little and then
joining together end to end. Examples of fibers are shown in Figs. 35
and 85. Speaking in general terms,
we may say that the whole body consists of tiny cells, either rounded and
thick, flat and thin, or elongated to
form fibers. Just as a wall is built
of distinct bricks or stones, so an
organ is made up of a number of cells.
All the solid parts of the body are
either cells or fibers which have grown

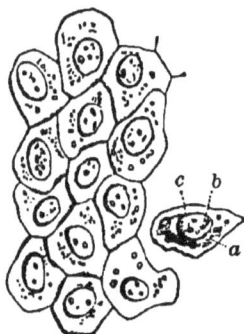

FIG. 6.—Flat cells from the surface of the lining membrane of the abdomen; *a*, cell body; *b*, nucleus; *c*, nucleoli.

from cells, except something which corresponds pretty
closely to the mortar which lies between the bricks of a
wall and holds them together. This latter material,
known in the body as *intercellular substance,* is in some
places abundant, in others scanty or absent.

Wherever found, the intercellular substance is made
by the cells which lie imbedded in it ; they pass it out
from their surfaces and repair it when necessary, and in
this respect it differs very essentially from the mortar
which a mason lays between his bricks.

Summary.—Cells are thus at bottom the things which

What are fibers ? How are they made ? In what respect may an
organ be compared to a brick wall ? What corresponds to the mortar
of the wall ? What makes the intercellular substance ? How ?
State briefly the relationship of its cells to the structure and working of the body.

make up the body and do its work; their forms and the way they are arranged together determine the form of the organs; the things which the kinds of cells found in it can do, determine the faculties of each organ. Some cells can make a great deal of hard intercellular substance, and are employed to construct the skeleton; others can change their shape and are used to form the organs which move the body; others can elaborate peculiar solvent liquids and are used in the organs of digestion; and so on through all the parts. Anatomy in the long run is a study of the forms which cells and intercellular substances may assume; and physiology a study of what the cells and intercellular substances of the body can do.

The physiological division of labor.—In a tribe of wandering savages, living by the chase, we find that each man has no special occupation of his own; he collects his own food, provides his own shelter, defends himself from wild beasts and his fellow-men. In a civilized nation, on the contrary, we find that most men have some one particular business: farmers raise crops and cattle; cooks prepare food; tailors make clothes; and policemen and soldiers protect the property and lives of the rest of the community; in other words, we find a *division of labor.* Just as the more minute the division of employments in it is, the more advanced a nation is in civilization, so an animal is higher or lower just as the duties necessary for maintaining its existence are distributed among different

What determines the form of an organ? What its faculties? Give examples of the employment of cells with different powers to do different things. What is Anatomy really? What Physiology? Explain what is meant by the physiological division of labor. In what class of nations is the division most minute? What decides whether an animal is higher or lower than another?

2

tissues and organs. In the lowest animals every cell is concerned in feeling, and moving, and catching food, and digesting, and breathing; in higher animals different cells are set apart in different organs for the execution of each of these separate functions.

Results of a division of labor.—From the division of employments in advanced communities, several important consequences result. In the first place, when every one devotes his time mainly to one kind of work, all kinds of work are better done : the man who always makes boots becomes much more expert than the man who is engaged on other things also : he can not only make more boots in a given time, but he can make better boots; and so in other cases. In the second place, when various employments are distributed among different persons there arises a necessity for a new kind of industry in order to convey that part of the special produce of any given individual which may be in excess of the needs of himself and his family to others who may want it, and to bring him in return such of their excess production as he may need. The conveyance of food from the country to cities, and of manufactures to agricultural districts, are examples of this sort of labor in civilized communities. Finally, there is developed a necessity for arrangements by which, at any given time, the activity of individuals shall be regulated in accordance with the wants of the whole community or of the world at large. This sort of regulation is still

What do we find all cells doing in the lowest animals? How do higher animals differ in this respect?

How does a division of labor influence the quantity of work done by a man? How the quality? Illustrate. What new kinds of employment arise when a division of labor becomes developed in a nation? Illustrate.

very imperfectly carried out even in the most advanced communities, and we accordingly hear from time to time of the over-production of this or that article ; but it is in part effected through the agency of capitalists who control the activities of many individuals in accordance with what they think to be the quantity of various articles likely to be required from time to time.

Exactly similar phenomena result from the division of physiological labor in the human body. Each tissue and organ doing one special work for the whole body, and relying on the others for their aid in turn, every sort of necessary work is better performed ; the tissue or organ, having nothing else to look after, is constructed with reference only to its own particular duty, and is capable of doing it extremely well. This, however, necessitates a *distributing mechanism* by which the excess products, if any, of the various organs, shall be carried to others which require them ; and a *regulating mechanism* by which the activities of each shall be controlled in accordance with the needs of the whole body at the time being. We accordingly find a set of organs, the *heart and blood-vessels*, which carry blood from place to place all over the body, the blood getting in its course something from and giving something to each organ it flows through ; and a set of *nervous organs* which ramify in every direction and regulate the activity of all the more important parts.

The chemical composition of the body.—If we go beyond the tissues to seek the ultimate constituents of the body, we must lay aside the microscope, and call in chem-

How does the division of duties in it affect the human body? What is the object of the distributing mechanism? What of the regulating? What are the distributing organs in the body called? What is the object of the nerve organs?

istry to our aid, to discover what elements and compounds make up the cells and intercellular substance.

Elements found in the body.—Of the many elements discovered by chemists, only sixteen have been found in the healthy human body.* Very few exist in it uncombined. Some oxygen is dissolved in the blood; and that gas is also found, mixed with nitrogen, in the lungs.

Chemical compounds existing in the body.—These are so numerous that it would be a long task to enumerate them, but some require mention : they may be divided into organic and inorganic. In a general way we may say that the organic constituents of the body, if all water were separated from them, would burn if put in a fire; and the inorganic components could not be made to burn.

Inorganic constituents of the body.—Of the inorganic constituents of the body, *water* and *common salt* are the most important; they are found in all the organs and liquids of the body. Phosphate and carbonate of lime are found in large proportions in the bones and teeth; and free hydrochloric acid (spirits of salts or muriatic acid) may always be found in healthy gastric juice, which dissolves some kinds of food in the stomach.

Organic constituents of the body.—All organic constituents of the body contain carbon, hydrogen, and

How many chemical elements exist in the body? Are most of them free or combined with others? Name those found free.

Into what main groups may be divided the chemical compounds existing in the body? Name some of the chief inorganic compounds helping to build the body. Where are they found in it?

What do all organic substances in the body contain?

---

* The elements found in the body in health are—carbon, hydrogen, nitrogen, oxygen, sulphur, phosphorus, chlorine, fluorine, silicon, sodium, potassium, lithium, calcium, magnesium, iron, and manganese.

oxygen; some contain nitrogen also. There are three chief kinds of them, viz.: albumens, fats, and carbohydrates.

Albuminous or proteid substances.—These are by far the most characteristic organic compounds existing in the body; they are only known as obtained from living beings, having never yet been artificially constructed in the laboratory; a good example is found in the white of an egg, which consists chiefly of albumen dissolved in water. All the tissues of the body which have any marked physiological property contain some albuminous substance, only such things as hairs, nails, and teeth being devoid of them. All albuminous bodies contain nitrogen, carbon, hydrogen, and oxygen; most of them sulphur and phosphorus in addition. The more important ones found in the body are, (1) *Serum albumen*, very like egg albumen, and found dissolved in the blood, along with another named (2) *fibrinogen*, which turns into (3) *fibrin*, when blood clots; (4) *Myosin*, found in the muscles and "setting" or coagulating after death, when it causes the death stiffening; (5) *Casein*, found in milk, and forming the main bulk of cheese.

Fats belong to the organic compounds in the body which contain no nitrogen; they consist solely of carbon, hydrogen, and oxygen. The chief fats in the body are *palmatin, stearin,* and *olein;* by proper treatment each can be split up into *glycerine* and a fatty acid; *palmitic, stearic,* or *oleic acid* as the case may be.

What is found in addition in some of them? How many chief varieties of organic compounds are there in the body? Name them. Give another name for albuminous substances. Can they be made artificially? Give an example of an albumen. What elements do albumens contain? Name the more important albumens of the body. Where are they found?
What elements do fats contain? Name the chief fats of the body. Into what may they be decomposed?

The carbohydrates also consist entirely of carbon, hydrogen, and oxygen; they belong to the same class of substances as starch and sugar. The most important carbohydrate in the body is *glycogen*, a sort of starch found stored up in the liver and muscles. *Glucose* or *grape sugar* also exists in the body; and *lactose* or *milk sugar* is found in milk.

What elements do carbohydrates contain? Name the most important found in the body? Where is glycogen found? Where milk sugar?

CHAPTER III.

PUTREFACTION, FERMENTATION, DISTILLATION, AND THE NATURE OF ALCOHOL.

Putrefaction.—It is within the experience of every one that most dead animal and many dead vegetable matters rapidly decompose if kept in a warm, moist condition; and that while decomposing they emit ill-smelling and unwholesome vapors. One of the problems of large cities is how to efficiently and economically remove the "garbage," which, if allowed to accumulate and decompose, would poison the community.

This decomposition, named putrefaction, is due to the activity of extremely minute living things which, wafted everywhere in the air, sooner or later settle on animal or vegetable remains and destroy them. The living things which cause putrefaction are known as Bacteria. Other kinds of Bacteria are the causes of most infectious diseases. Many have the faculty of lying dormant in a dry condition for months or years. In this state they form light dust, capable of transference by the softest breeze; but as soon as they settle on a suitable material they become active, and multiply with great rapidity.

The Bacteria of putrefaction are among those which can live in the dry state for some time ; and they always form

Give some examples of putrefaction. To what is it due?

23

part of the dust floating in the air; so it is easy to understand why fresh meat or fish left exposed to the air soon putrefies. As the Bacteria multiply much more rapidly in warm than in cold weather it is harder to keep meat in summer than in winter.

How Meats and Vegetables are preserved from Putrefaction.—Man has discovered various methods of preserving for a long time foods liable to putrefy. The simplest method is to freeze the food or at least keep it very cold, a principle applied on a large scale in the refrigerator cars which convey meat from Western States to cities in the East. On a smaller scale this plan is used in the refrigerator found in most homes.

Another simple method of preserving meat and fish from putrefaction is to dry them rapidly and keep them dry, since the Bacteria of putrefaction thrive only in the presence of moisture: but the most common methods of preserving foods are two; one of which depends on the killing of all Bacteria present and preventing the entry of fresh ones, the other on adding substances which prevent the growth of the putrefactive organisms.

All the great "canning" industries of the United States depend on the fact that putrefactive Bacteria are killed by exposure to the temperature of boiling water; and that cans of meat and vegetables so exposed can be sealed while hot, so as to prevent the entry of living Bacteria.

Well-known instances of the use of substances which prevent the development of putrefactive organisms are the

How is it that Bacteria are found in dust? Why is it harder to keep meat in summer?

Describe a plan of preserving meat from decomposition; a second plan; a third; a fourth.

employment of salt to preserve meats and fish; and of strong vinegar or plenty of sugar in the case of pickles and preserves.

Fermentation.—Many microscopic plants, some of them Bacteria and some not, decompose complex dead organic matters into simpler ones without any accompanying emission of foul smells: when this is the case the process is usually called a fermentation instead of a putrefaction. There is, however, no real scientific distinction between a putrefaction and a fermentation: whether the products smell well or ill is a very secondary matter; the essential fact is the destruction, by certain kinds of minute living things, of the complex and elaborated products of other plants or animals. For this reason biologists make no distinction between putrefactions and fermentations: they name all such processes *fermentations.*

Formed and Unformed Ferments.—It is still usual to speak of certain dead substances as ferments; for example, the substance (ptyalin) in our saliva which turns starch into sugar (p. 197), and the pepsin (p. 177) of the gastric juice, are frequently called ferments. It is very desirable to draw a sharp line between such dead things as pepsin (which can be chemically extracted from the lining of a dead stomach and still act in breaking down or changing organic substances) and such living things as Bacteria, causing fermentations as a side result of their own life-actions, and quite inactive when killed.

It is now customary among physiologists to distinguish

What distinction is made in ordinary language between a putrefaction and a fermentation? What is the essential nature of both processes? What is an "unformed" ferment? Name one or two. How do such ferments differ from "formed" ferments?

the two classes of ferments, the living and the dead respectively, as *formed* and *unformed.**

Every living ferment has its own field of work. Some act on albumens and decompose them ; others break up fats and oils and make them rancid. Still others destroy starches and sugars: the souring of milk, for example, is due to the alteration of the milk-sugar by a ferment which turns it into an acid (lactic acid). It might, at first sight, seem that all true ferments were merely mischievous; for they destroy much to make little. But if they did not destroy dead animals and plants the earth would now be so covered by the dead bodies of its former inhabitants, that neither animal nor plant could live on it: the land would be covered by a layer of mummies. Ferment organisms break up all this dead and useless matter and resolve much of it into simpler substances, which, in turn, serve to nourish succeeding living things. For example, when animal or vegetable substances putrefy, much of their carbon is given off to the air as carbon dioxide gas, which green plants then use in making food for animals (see p. 132).

Alcoholic Fermentation.—Among the many fermentations there is one which demands special attention on account of its consequences to mankind: one of its products is the liquid known as alcohol—a poison which is the main constituent of all commonly used intoxicating drinks. A poison is any substance which tends when taken into the body to injure or kill it.

By very roundabout and expensive methods chemists can

State a use of fermentative organisms. What substance do all ordinarily used intoxicating drinks contain ? What living ferment produces alcohol ? What is a poison ?

---

* Another very useful name now coming into use is that of *enzyme* for a dead ferment such as pepsin, the name ferment being restricted to formed ferments.

make alcohol without the aid of a ferment; but in so far as alcoholic drinks are to be found in the bar-room or drug-store for sale, they have all (whether beer, cider, ale, por-ter, wine, whiskey, brandy, cordials, or "tonics") been made by the action of ferments known as yeast. The sub-stance commonly known as bakers' or brewers' yeast is only one kind of the yeast or ferment which is the cause of alcoholic fermentation.*

The yeast plant cannot thrive in strong solutions of sugar, and therefore preserves and syrups may be kept for a long time without fermentation. But in weak solutions of sugar the yeast plant thrives and turns the sugar into carbon dioxide and alcohol.†

Alcoholic Fermentation a Waste.—Students who have some knowledge of chemistry will see at once that the breaking down of a complex organic substance like sugar into simpler ones as alcohol and carbon dioxide means a loss or waste of energy or working power. Yeast while splitting up the sugar takes to itself or gives off as heat a quantity of the force which was formerly stored in the sugar. Indeed, so much heat is given off during alcoholic fermentation that in breweries large quantities of ice have to be used to cool the fermenting liquid, or it would get so warm that the fermentation would be stopped. So far as man's food is concerned, fermentation that produces

Can alcohol be made without the aid of the ferment? How are preserves protected from fermentation? Why is alcoholic fer-mentation a waste of energy? Why is it, so far as man's food is concerned, worse than a waste?

---

*The technical or generic name of the yeast plant or alcoholic ferment is *Saccharomyces.* There are several species of this genus: the yeast used by the baker and brewer is *Saccharomyces Cerevisiæ.*

† Small quantities of other substances are also found in alcoholic fermenta-tion, and a small part of the sugar is used as food by the yeast.

alcoholic liquors is pure waste. It destroys sugar, which is a useful food, and gives instead carbon dioxide gas, which is useless to him, and alcohol, which is worse than useless as an article of diet, though having many uses in the arts, as in making varnishes. Sugar is a useful nourishing food; alcohol is a poison which tends to injure nearly all the organs of the body.

Common Yeast, as usually seen, is a thick yellow-brown liquid. Examined with a microscope, this liquid is found

to contain myriads of tiny plants belonging to the group of Fungi. These plants are ovoid, and may be single or grouped together. In suitable liquids they multiply

Yeast Plants: magnified greatly (about 800 diameters).

with great rapidity by young cells budding off from the older ones. Yeast cells may be dried and yet retain their vitality for a long time, ready to set up alcoholic fermentation whenever they reach a dilute solution of sugar.

The Process of Alcoholic Fermentation may be illustrated by brewing. To make beer, brewers usually commence with barley, which contains much starch. Yeast will not act directly on starch, so the grain is kept moist and warm until it sprouts; during this process the greater part of the starch is turned into sugar ready to nourish the young plant. Before the plant uses the sugar the grain is heated until it is killed: the product is known as malt. The malt is crushed, and the sugar and some other things dissolved out by water; this solution is *beerwort,* a sweetish liquid.

Name some real use of alcohol. Describe the yeast plant. How is it that dilute sugar solutions exposed to the air are apt to undergo alcoholic fermentation? How is malt made? Why is barley malted before used for brewing? What is wort? How made?

To this the brewer adds yeast, and very soon a remarkable series of phenomena occurs. Bubbles of gas are given off abundantly, so that the whole liquid becomes frothy; this gas is carbon dioxide, useless to animals either as food or for breathing. As the fermentation proceeds, there is less gas because most of the sugar has been destroyed and the yeast has little more to break up, but alcohol has accumulated in the liquid. Hops and other flavoring matters may have been added to the wort during its fermentation, but this does not essentially influence the process, which is the destruction of sugar and the formation of alcohol and carbon dioxide, under the influence of the yeast fungus.

When the sugar of the wort is almost destroyed, we find the bubbling of gas to nearly cease. The sweet-tasting wort has lost its sweetness, the yeast settles to the bottom of the vessel, and the product is the intoxicating liquid known as beer, and containing alcohol instead of the original sugar. During the fermentation good food has been destroyed and a dangerous liquor has been produced.

Some other Alcoholic Fermentations.—The fermentation of barley wort by yeast is but one example of many similar cases. The juices of most fruits and vegetables contain a sugar capable of alcoholic fermentation. As the fruits ripen the germs of various alcoholic ferments collect upon their surfaces and stems, from whence they are easily carried into the juice as it is pressed out. They may also be found in a dry state floating about in the air in the vicinity of ripening fruits; but they are never found in the juice itself while it is in the fruit. Hence alcoholic fermentation never starts in the interior of fruits with unbroken skins.

What are the essentials in the process of brewing? How is the wort changed when the fermentation is nearly completed? Why are most vegetable and fruit juices fermentable?

When the juice of apples is pressed out and exposed to the air, ferments from the surface of the fruit, or floating in the dust of the air, reach it, and break up its sugar into carbon dioxide and alcohol; the result is cider, an intoxicating liquor. Similarly with pear-juice, which when fermented gives perry. In like manner, the ferments that rest upon the skins and stems of grapes are easily carried into the fermenting vats * when the grape-juice is exposed, and seizing upon the sugar, produce the alcohol-containing liquor called wine. Wines are also made from other fruits, such as currants, raspberries, elderberries, etc. It is important to remember in this connection that the character of a substance does not depend upon its quantity, but upon its quality. All beers, ales, porters, ciders, and wines (home-made or not) contain alcohol in varying quantity, but the character of the alcohol is in every case the same.

Alcoholic Appetite.—One of the worst features of the poisonous characteristics of alcohol is its power even in small quantities to create a craving for itself that often becomes irresistible. It is therefore the nature of beer, cider, and wine to lead to an increasing use of alcohol. A continued increasing use of alcohol results in drunkenness. The craving for alcohol is not a natural appetite; it is not a demand set up by the tissues of the body for a new supply of material needed for construction or repair. Enough food to supply this demand satisfies, more causes satiety;

What is cider? Why does exposed apple-juice ferment? What is wine? Does the nature of a substance depend upon its quality or quantity? What substance do all beer, ale, cider, perry, and wines contain? What can you say of the character of alcohol in all these? What is one of the worst characteristics of alcohol, and what danger therefore attends the use of the lighter liquors?

---

* The yeast found most frequently on the surface of ripe grapes is the *Saccharomyces ellipsoidius,* called also the *ferment of wine.*

but the craving for alcohol increases with its supply. The only remedy is to cease the use of the liquor craved.

The difficulty of overcoming the alcoholic appetite once it has been acquired, and the danger of arousing it again after it has long been successfully resisted, make it unwise to use alcoholic liquors for flavoring food.

There are many Alcohols.—When yeast causes fermentation of dilute sugar solutions, other substances than carbon dioxide and common alcohol are formed in small quantity. Many alcohols are known to chemists, but when the word alcohol is used without qualification, common or ethyl alcohol is understood. The hurtful actions on the human body of ethyl alcohol are stated in other chapters of this book, and are serious enough. But in most fermentations of sugary liquids under the influence of yeast other alcohols are also produced, much more poisonous than common alcohol. The most important of these specially poisonous alcohols is amylic alcohol or fusel-oil. Few kinds of spirituous drinks are free from it, and it is a very potent poison.

Distillation.—To obtain liquids containing more alcohol and less water than that found in fermented liquors, the process named *distillation* is employed.

Alcohol boils and passes off as vapor at a temperature much below that at which water is converted into steam. Consequently, if a solution of a small proportion of alcohol in a large quantity of water be heated (for example, beer or wine), the alcohol will evaporate before much of the water does, and can be condensed and collected in a cold receptacle,

How does the craving for alcohol differ from the appetite for food? What is the only remedy for the alcoholic appetite? What is the name of common alcohol? What is fusel-oil? How do distilled differ from fermented liquors? How does distillation bring about a greater proportion of alcohol in the distilled liquid?

mixed with but little water. This fact is used in preparing such liquors as whiskey and brandy. The fermented extract of rye or corn, or fermented grape-juice, is heated, and the first products of the evaporation, which contain much alcohol, are collected in a cold vessel. Some water and some flavoring matters (varying with the fermented substance) come over with the alcohol, and influence not only the flavor but the intoxicating effect of various distilled liquors.

Brandy, whiskey, gin, and rum all contain more alcohol than an equal bulk of simply fermented liquors, such as wine, beer, or ale; but all such liquors, whether merely fermented, or fermented and then distilled, contain alcohol, and should never be used as a food. They are poisonous beverages, and injurious to health. Distilled liquors contain more alcohol than fermented, and are therefore more mischievous.

Alcohol, or, in chemical phraseology, *ethyl alcohol* (C_2H_5, HO), is a clear liquid, lighter than water and very inflammable. It has a great avidity for water, and even when diluted, as in whiskeys or brandies or spirits of wine, will extract water from animal tissues and make them hard. To this fact, illustrated in every museum of natural history, is due part of the harm wrought by alcohol upon living human tissues.

To obtain pure ethyl alcohol is a difficult matter on account of its great tendency to take water from the atmosphere or any other water-containing solid, liquid, or gas. Pure ethyl alcohol is only to be found in chemical labora-

---

Why do distilled liquors vary in flavor? How do fermented drinks differ from distilled drinks? What is the substance common to all of them? What are its effects upon health?

tories; diluted alcohol may be found in every bar-room. Pure or diluted, adulterated, or mixed with other alcohols such as fusel-oil, it is a poisonous substance, harmful to healthy persons. Its hurtful influence on the joints, muscles, heart, digestive organs, blood-vessels, and nervous system are indicated in other pages of this book; while its influence in weakening the will and character can hardly be exaggerated.

Describe pure ethyl alcohol. How does it act on animal tissues even when diluted? Why is it hard to obtain pure ethyl alcohol? Where may impure ethyl alcohol be always obtained? Name some organs which alcoholic drinks injure?

CHAPTER IV.

THE SKELETON.

The skeleton * of the human body is composed of three materials : *bone, cartilage,* and *connective tissue.*

The bones form the main supporting framework of the body, and determine its shape ; they provide levers on which the muscles moving the body pull, and are arranged so as to surround cavities in which soft, delicate organs, as brain, spinal cord, and heart, may lie in safety.

Cartilage finishes off many bones at joints, forming elastic pads with smooth surfaces ; it is also used instead of bones in some parts of the skeleton where considerable flexibility is required. Cartilage affords one of the best tissues of the body for the examination of intercellular substance. A thin slice of it highly magnified, Fig. 7, shows the *cartilage cells, a, b,* scattered through an almost structureless material. Very young cartilage consists of

Of what is the skeleton made up ? What functions do the bones fulfill ? Where is cartilage found ? What are its purposes ? What is seen when a thin slice of cartilage is highly magnified ? Of what does young cartilage consist.

---

* There are two kinds of skeleton met with in the animal kingdom ; the external skeleton or *exoskeleton,* and the internal skeleton or *endoskeleton.* The exoskeleton is made by the skin, either in it or on it ; examples are found in the shells of clams and lobsters · the scales of fishes and snakes ; the tortoise-shell of turtles ; the feathers of birds; the hair and claws of beasts. In man the exoskeleton is only slightly developed; hair, nails and teeth belong to it.

the cells only, but these lay down around them more and
intercellular substance, until at last it forms the main bulk

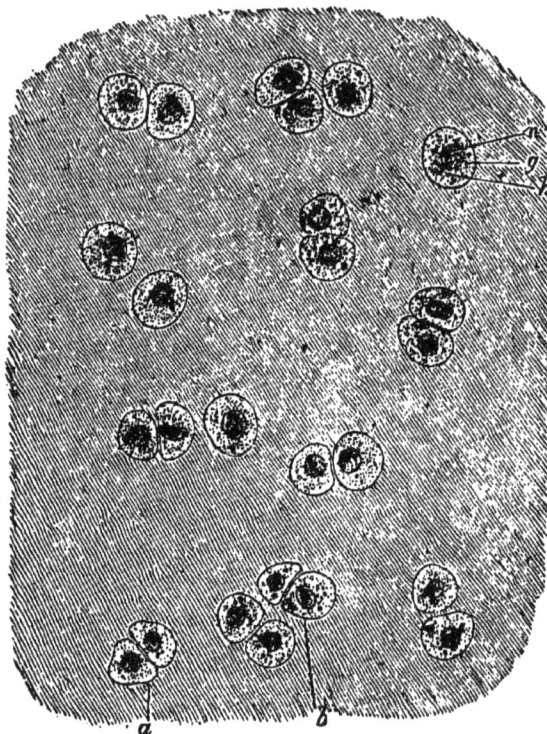

Fig. 7.—A thin slice of cartilage highly magnified.

of the cartilage, and gives this the elasticity and flexibility
for which it is used in the body.

Connective Tissue occurs partly in the form of stout
cords—*ligaments*—which bind different bones together;
or which, called *tendons*, attach muscles to bones. It
also supplements the coarser bony skeleton by a finer
one, which extends as a fine network through all the soft
parts of the body, making a sort of microscopic skeleton

What is a ligament? A tendon? Of what are ligaments and
tendons composed? Where else do we find connective tissue?

for its cells, and being laid down as a soft packing material, a good deal like raw cotton, in the crevices between different organs, as shown at *s*, Fig. 3, where it is seen between the muscles of the fore-arm. This tissue is, in fact, so widely spread over the body, from the skin outside to the lining of the alimentary canal within, that if we could employ a solvent which would corrode away all the rest, and leave only the connective tissue, a very perfect model of all the organs would be left; something like a skeleton leaf, but far more minute in its tracery.

Articulations and Joints.—If the pieces forming the hard frame-

For what purposes ? What would be left if all materials composing the body except its connective tissue were dissolved away ?

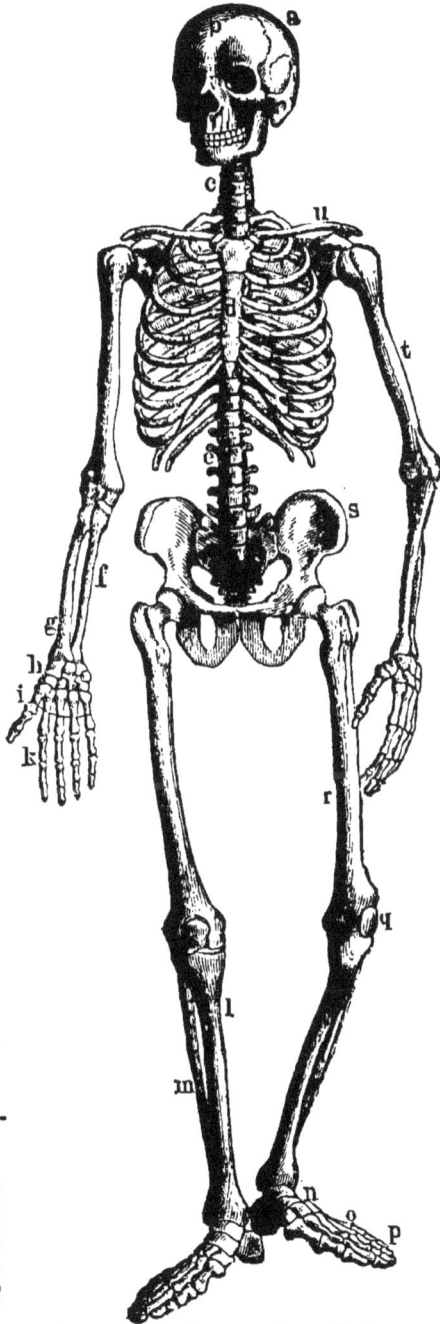

FIG. 8.—The bony and cartilaginous skeleton.

work of the body were put together like the beams and planks of a frame house, the whole mass would be rigid and immovable ; we could not raise a hand to the mouth, or put one foot before another. In order to attain mobility the bony skeleton is made up of more than two hundred separate pieces, joined together; the points where they meet are called *articulations.* An articulation at which a considerable range of movement is permitted is called a *joint.* The ends of bones which rub over one another in a joint are always covered by a very smooth layer of cartilage.

The bony skeleton (Fig. 8) consists of an *axial skeleton,* supporting head, neck, and trunk, and an *appendicular skeleton,* supporting the limbs and attaching them to the trunk.

The Axial Skeleton.—The fundamental portion of this is the *backbone, spinal column,* or *spine,* partly seen at *e* and *c,* Fig. 8, and represented isolated from the rest of the bones and viewed from the left side in Fig. 9. It forms an axis, on which the rest of the body is carried. On the upper end of the vertebral column is the *skull, a, b,* Fig. 8, and attached by ligaments to the under surface of the skull is the *hyoid bone,* to which the root of the tongue is fastened.

Attached to the sides of part of the spine are the dorsal ends of the *ribs,* slender bones which curve round the

Why is the skeleton made up of separate bones ? What are articulations ? What is a joint? What covers the end of a bone in a joint ?

What are the chief divisions of the bony skeleton ? What parts of the body does the axial skeleton support ? What is the appendicular skeleton?

What is the fundamental portion of the axial skeleton? What does it bear on its upper end? What is the hyoid bone ?

What are the ribs? Where are their ends fixed ?

sides of the chest and are united in front to the *sternum,* or *breastbone, d,* Fig. 8.

Skull, hyoid bone, vertebral column, ribs, and sternum together form the axial skeleton.

The appendicular skeleton consists of the *pectoral* and *pelvic girdles,* attaching the limb bones to the axial skeleton ; and of the bones of the limbs themselves.

The pectoral arch or girdle consists on each side of a *clavicle* or *collar-bone, u,* and a *scapula* or *shoulder-blade,* which latter is a flat, triangular bone, lying on the back of the chest outside the ribs. The clavicle is a slender curved bone like an italic *f* in form. Its outer end is attached to the scapula, and its inner end to the top of the sternum. It serves to brace out and support the shoulder-joint, and to prevent it from falling downwards and inwards toward the front of the chest. It is absent in beasts which use their fore limbs for walking only, as horses, dogs, and cattle, but is well developed in monkeys and bats.

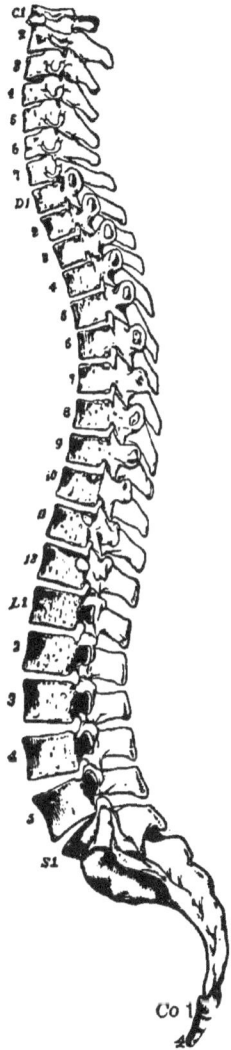

FIG. 9.—Side view of the spinal column.

Name the parts which make up the axial skeleton?

Of what main divisions does the appendicular skeleton consist? What bones exist in the pectoral arch? Where does each lie? What is its shape? Name some animals which have no collar bones, and some which have them.

The skeleton of the upper limb consists of : (1) The arm bone, or *humerus, t,* Fig. 8, which extends from the shoulder to the elbow, and meets the scapula at the shoulder-joint ; (2) of two forearm bones, the *radius, g,* and *ulna, f,* the radius being on the thumb side ; and (3) of twenty-seven hand bones. Of the hand bones eight, the *carpal bones, h,* lie in the wrist ; five, the *metacarpal bones, i,* in the palm of the hand ; and fourteen, the *phalanges, k,* in the thumb and fingers—two for the thumb, and three for each finger.

The pelvic arch or girdle consists of a single bone, the *os innominatum, s,* on each side ; this is firmly fixed at its dorsal end to the lower part of the back-bone, meets its fellow ventrally at the lower end of the abdomen, and bears a deep socket on its outer side, into which the upper end of the thigh-bone fits.

The skeleton of the lower limb consists of : (1) The *thigh-bone* or *femur, r,* the longest bone in the body, bearing above a hemispherical knob fitting into a hollow on the outside of the os innominatum, with which it forms the hip-joint ; (2) of two bones, *tibia* and *fibula, l* and *m,* in the lower leg, the former on the great toe side ; (3) of

How many bones are there in the upper limb? Which of these lies between the elbow and shoulder joints? With what bone does it articulate above? Name the forearm bones. Which of them lies to the outer side when the palm of the hand is turned forward? How many bones are there in the hand? Where are the carpal bones ; How many of them are there? The metacarpal bones ; position and number? The phalanges of the hand ; position and number ?

How many bones form the pelvic arch? What are they named? To what are their dorsal ends attached? Where do they meet one another ? What bone of the leg forms a joint with the innominate bone ?

What is the first bone in the skeleton of the lower limb ? How does it end above ? With what bones does it articulate ? At what joints ? What bones extend from the knee to the ankle joint ? Which of them is on the inner side of the leg ?

the knee-cap or *patella, q,* in front of the knee-joint ; (4)
of twenty-six foot bones. Of the foot bones seven, the
tarsal bones, n, lie below the ankle-joint ; five, the *meta.
tarsal bones, o,* succeed these in the front half of the sole

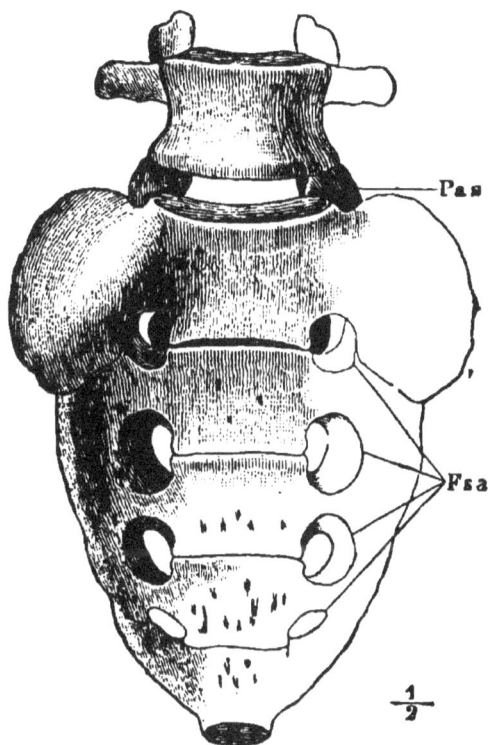

Fig. 10.—The last lumbar vertebra and the sacrum seen from the ventral side.
Fsa, anterior sacral foramina.

of the foot ; and fourteen *phalanges, p,* are found in the
toes ; two in the great toe and three in each of the others.

Where is the patella?   How many bones are there in each foot?
Into what groups are the foot bones classified?   Where are the
tarsal bones? How many ?   The metatarsal ?   How many ?   The
phalanges ?   How many ?

The vertebral column.—(Fig. 9.) The upper portion of the spine consists of twenty-four separate bones, each called a *vertebra;* these are piled one above the other, and separated by elastic pads made of cartilage and connective tissue. Seven vertebræ (*cervical, C* 1–7) are found in the neck ; twelve (*dorsal, D* 1–12) lie at the back of the chest and carry the ribs ; and five (*lumbar, L* 1–5) are in the loins.

Below the separate vertebræ comes the *sacrum,* (*S* 1), which is shown as seen from its ventral aspect in Fig. 10, along with the lowest lumbar vertebra. In childhood the sacrum consists of five distinct vertebræ, but these grow together afterwards, though cross ridges remain indicating the original lines of separation. Succeeding the sacrum and forming the lower end of the spine is the *coccyx* (*Co,* 1–4, Fig. 9), a single bone in adults, though consisting of four pieces in children.

The structure of a vertebra.—Those vertebræ which remain permanently separate resemble one another in general form, with the exception of the uppermost two. As an example we may take the eleventh from the skull, that is the fourth dorsal vertebra (Figs. 11 and 12).

In it we find (1) a thick bony mass, *C,* rounded on the sides and flattened above and below where it is turned toward its neighbors ; this part is the *centrum* or *body* of

Of what is the upper portion of the backbone composed? What are the bones forming it called? What lies between them? How many vertebræ in the neck ? In the chest region? In the loins?

Of what parts is the lower portion of the vertebral column composed? How many vertebræ form the sacrum? At what period of life are they separate? How is this original separation indicated on the sacrum of adults? How many vertebræ are united to form the coccyx?

What vertebræ differ essentially in form from the rest? Describe a typical vertebra.

the vertebra; the series of vertebral bodies forms the bony partition (*e, e,* Fig. 1) already mentioned as existing in the trunk between the neural and hæmal cavities. (2) An arch attached to the dorsal side of the centrum; it is the *neural arch,* *A,* and with the centrum incloses the neural ring (*Fv*). The vertebræ being piled one above the other the successive neural rings form the neural tube, in the

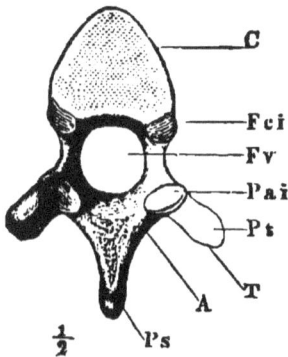

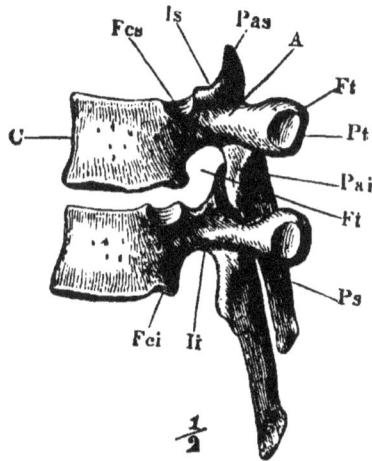

FIG. 11. FIG. 12.

FIG. 11.—A dorsal vertebra seen from behind, *i.e.,* the end turned from the head.

FIG. 12.—Two dorsal vertebræ viewed from the left side, and in their natural relative positions. *C,* the body; *A,* neural arch; *Fv,* the neural ring; *Ps,* spinous process; *Pas,* anterior articular process: *Pal,* posterior articular process: *Pt,* transverse process; *Ft,* facet for articulation with the tubercle of a rib; *Fcs, Fci,* articular surfaces on the centrum for articulation with a rib.

cavity of which the spinal cord lies. (3) Projecting from the body and arch are several processes; one reaching out from the dorsal side of the arch is the *spinous process;* the row of spinous processes which may be felt through the skin along the middle of the back has given the name of spinal column to the whole backbone.

What constitutes the hard partition between the dorsal and ventral cavities of the trunk? How is the neural tube formed? Why is the spinal column so named?

Where the arch joins the centrum it is narrowed to a stalk or *pedicle, Ii,* Fig. 12. When the vertebræ are placed together in their natural relative positions, apertures (*Fi*), leading into the neural canal, are left between their narrower portions; through these apertures (called the *intervertebral foramina*) nerves pass out from the spinal cord.

The atlas and axis.—The first and second cervical vertebræ differ considerably from the others. The first, called the *atlas* (Fig. 13), carries the head; it has a very

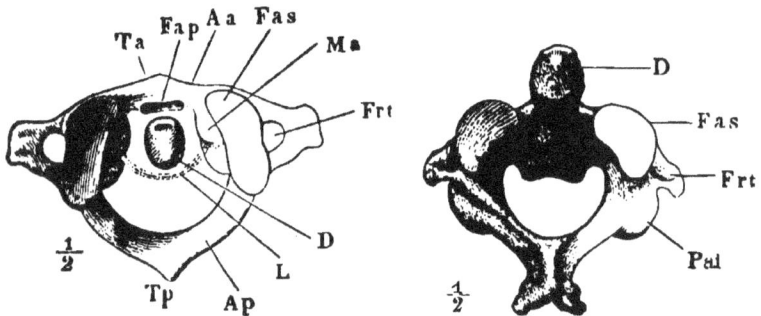

FIG. 13.
FIG. 14.

FIG. 13.—The *atlas.* FIG. 14.—The *axis. Aa,* body of atlas. *D,* odontoid process of axis; *Fas,* facet on upper side of atlas with which the skull articulates; and in Fig. 13, anterior articular surface of axis; *L,* transverse ligament; *Frt,* vertebral foramen.

small body and a very large neural ring. A ligament, L, divides the ring into a ventral and a dorsal portion; the spinal cord passes through the latter and a bony peg, D, lies in the former. The peg is the *odontoid or tooth-like process.* This reaches up from the second cervical or *axis vertebra* (Fig. 14) and forms a pivot around which the atlas,

How are the intervertebral foramina formed? What is their purpose?

What is the first cervical vertebra called? Describe its general form. How is its neural ring divided? What lies in each division? What is the second cervical vertebra named? What is the odontoid process? Around what pivot does the head rotate when the face is turned on either side?

carrying the skull with it, rotates when the head is turned from side to side.

On the anterior (upper) surface of the atlas are a pair of shallow hollows, *Fas;* into these fit a pair of knobs, found towards the back of the under surface of the skull (Fig. 20), which glide in the hollows during nodding movements of the head.

Uses of the mode of structure of the spinal column.— When the backbone is viewed from one side (Fig. 9) it is seen to present four curvatures; one in the neck, convex ventrally, is followed by a curve in the opposite direction in the dorsal region; in the loins the curvature is again convex ventrally, and opposite the sacrum and coccyx the reverse is the case. These curves add greatly to the springiness of the spine, and prevent the transmission of sudden jars along it.* The compressible elastic pads placed between the centra of the vertebræ promote the same end; the skull, containing the soft brain (which would be readily injured by mechanical violence) and the spinal cord, contained in the backbone itself, are thus protected from jarring in running, jumping, &c.

The compressible pads between the bodies of the vertebræ allow of a certain range of movement between each pair, so that the column as a whole may be bent to a con-

How is the skull articulated to the backbone?
How many curvatures are there in the backbone? What is their direction?
What results from the curvatures of the spinal column? What is the object of the pads between the vertebræ?

---

* Take a straight but tolerably flexible and elastic bar, as a lath, or, better still, a thin steel rod. Hold it vertical, with one end resting on the floor, and give a smart blow on the upper end; the jar will be sudden and violent. Now bend the rod and hit it again; the jar will be much less, as the curved rod yields somewhat to the blow on its top.

siderable extent in any direction. On the other hand, these pads so limit the movement that no sharp bend can occur at any one point, such as might tear or bruise the spinal cord lying in the neural canal.

The sacral vertebræ grow together firmly to give a solid support to the pelvic arch, which transmits the weight of all the rest of the body to the lower limbs when we stand.

Summary.—The back-bone is rigid enough to support all the rest of the body; flexible enough to bend considerably in any desired direction, yet not sharply at any one point; and elastic enough to destroy or greatly diminish any sudden jar or jerk which it may receive. It is one of the most beautiful pieces of mechanism in the body.

The ribs are twelve in number on each side (Fig. 15). They are slender curved bones embracing the sides of the chest, and attached at one end to the dorsal vertebræ. Ventrally each rib ends in a *costal cartilage;* the cartilages of the seven upper pairs are directly articulated to the sides of the breast-bone. The eighth, ninth, and tenth cartilages join those of the ribs above them; the eleventh and twelfth are not attached to the rest of the skeleton at their ventral ends, and are known as the *free* or *floating ribs*.

How is it that we can bend the backbone? How is the extent of bending at any one point limited? Why?

Why do the sacral vertebræ grow together?

State briefly the mechanical properties of the vertebral column.

How many ribs are there? What is their shape? To what are their dorsal ends attached? How does each rib end ventrally? To what are the costal cartilages of the first seven ribs attached? To what the next three costal cartilages? Which are the floating ribs? Why so called?

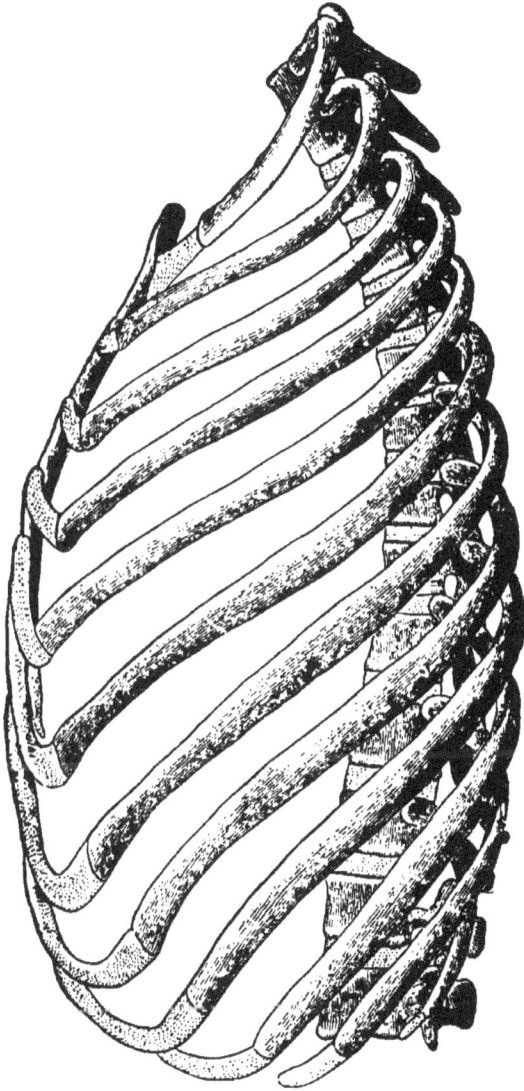

Fig. 15.—The ribs of the left side, with the dorsal and two lumbar vertebræ, the rib cartilages, and the sternum.

The skull (Fig. 16) is composed of twenty-eight bones : eight of these, forming the *cranium*, are so arranged as

How many bones in the skull?

to surround the brain and protect the ears ; six lie inside the ears ; and the remaining fourteen support the face,

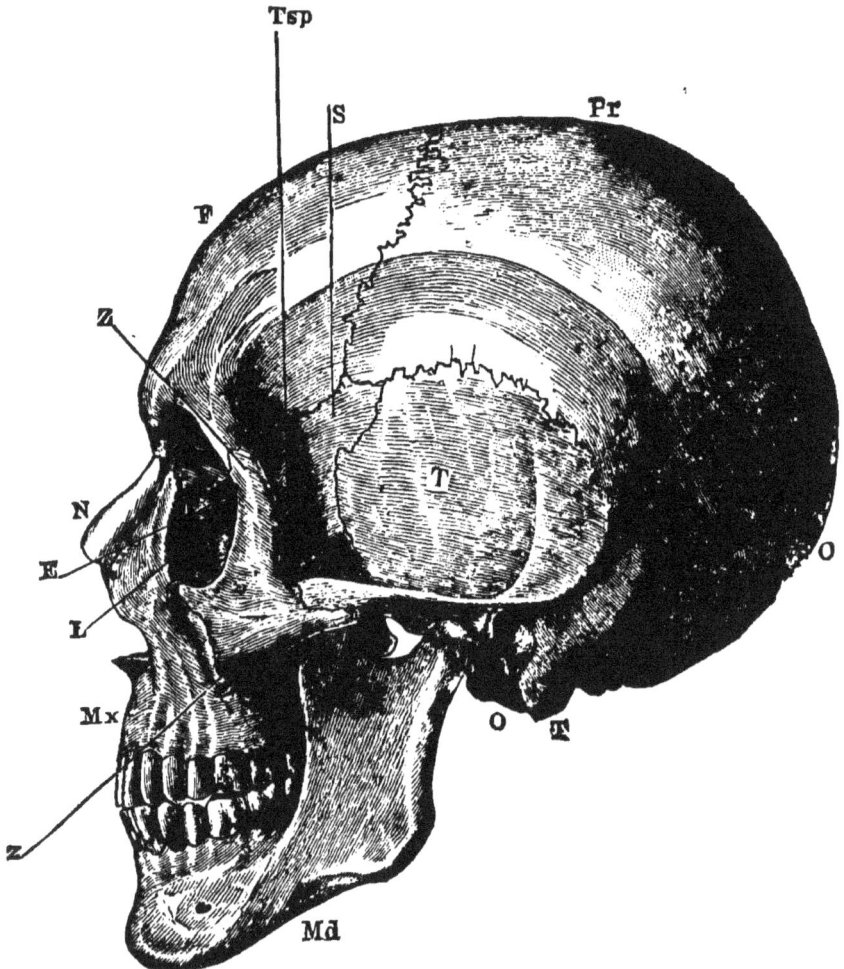

FIG. 16.—A side view of the skull. *O*, occipital bone ; *T*, temporal · *Pr*, parietal ; *F*, frontal ; *S*, sphenoid ; *Z*, malar ; *Mx*, maxilla ; *N*, nasal ; *E*, ethmoid ; *L*, lachrymal ; *Md*. inferior maxilla.

surround the mouth and nose, and (with the aid of some of the cranial bones) form the eye-sockets.

How many in the cranium? What purposes do the cranial bones fulfill? How many lie inside the ears? How many bones in the face? What parts do the face bones support and protect?

The cranium is a box with a thick floor (Fig. 1), continuing forwards the partition which in the trunk separates the neural from the hæmal cavity. On its under side (Fig. 20) are many small apertures through which nerves and blood-vessels pass in or out, and one larger one, the *foramen magnum*, through which the spinal cord passes in to join the brain.

The cranial bones (Fig. 16) are the following : 1. *The occipital bone*, O, unpaired, and having in it the foramen magnum. It lies at the back of the skull. 2. *The frontal bone*, F, also unpaired, lies in the forehead. 3. *The parietal bones*, Pr, two in number, meet one another above the middle of the crown of the head, and form a great part of the roof and sides of the skull. 4. *The temporal bones*, T, one on each side, opposite the temples ; on the exterior of each temporal bone is a large aperture leading into the ear cavity, which is contained in this bone. 5. *The sphenoid bone*, unpaired, and lying in the middle of the base of the skull, but sending out a wing, S, which reaches some way up each side, just in front of the temporal. 6. *The ethmoid bone*, E, forms the partition between the brain and nose chambers, and part of that between the nose and the eye socket.

The facial skeleton.—The majority of the face bones are in pairs, but two are single ; one of these is the *lower jaw bone* or *mandible*, Md, Fig. 16 ; the other is the *vomer*,

With what is the floor of the cranium continuous? Where does the cranium present holes through it? What are most of these apertures for? What is the largest aperture called? What enters the brain case through it?

Name the cranial bones. State where each lies. Through which does the foramen magnum pass? Which contains the ear chamber? Which of them are unpaired?

which forms part of the partition between the two nostrils.

The paired face bones are : 1. The *maxillæ* or *upper jaw-bones, Mx,* which carry the upper teeth and form most of the hard palate separating the mouth from the nose. 2. *The palate bones,* completing the bony palate, and behind which the nostril chambers communicate by the *posterior nares* (Fig. 20) with the throat cavity, so that air can pass in or out through them in breathing. 3. *The malar* or *cheek-bones, Z.* 4. *The nasal bones, N,* roofing in the upper part of the nose. 5. *The lachrymal* or *tear-bones, L,* small and thin, lying between the eye-socket and the nose. 6. *The inferior turbinate or spongy bones,* which lie inside the nose, one on the outer side of each nostril chamber.

The cranial sutures.—All the bones of the skull, except the lower jawbone, are immovably joined together. In the case of most of the cranial bones this occurs by a dove-tailing, like that used by cabinet-makers. Each bone has its edges notched, and the notches fit accurately into hollows on the bone it articulates with ; this kind of articulation is called a *suture ;* it is well seen in Fig. 16, between the parietal bone and those in front of, behind, and below it.

Comparison of the upper and lower limbs and their supporting arches. The bones of these have already

Name the unpaired face bones.

Where does each lie? Name the paired face bones. State the position of each in the skull. What bone carries the lower teeth? Which the upper? What bones form the hard palate? By what openings do the nose chambers communicate with the throat? Behind what bone do these openings lie?

What cranial bone is movable? How are most of the cranial bones joined together? Describe a suture?

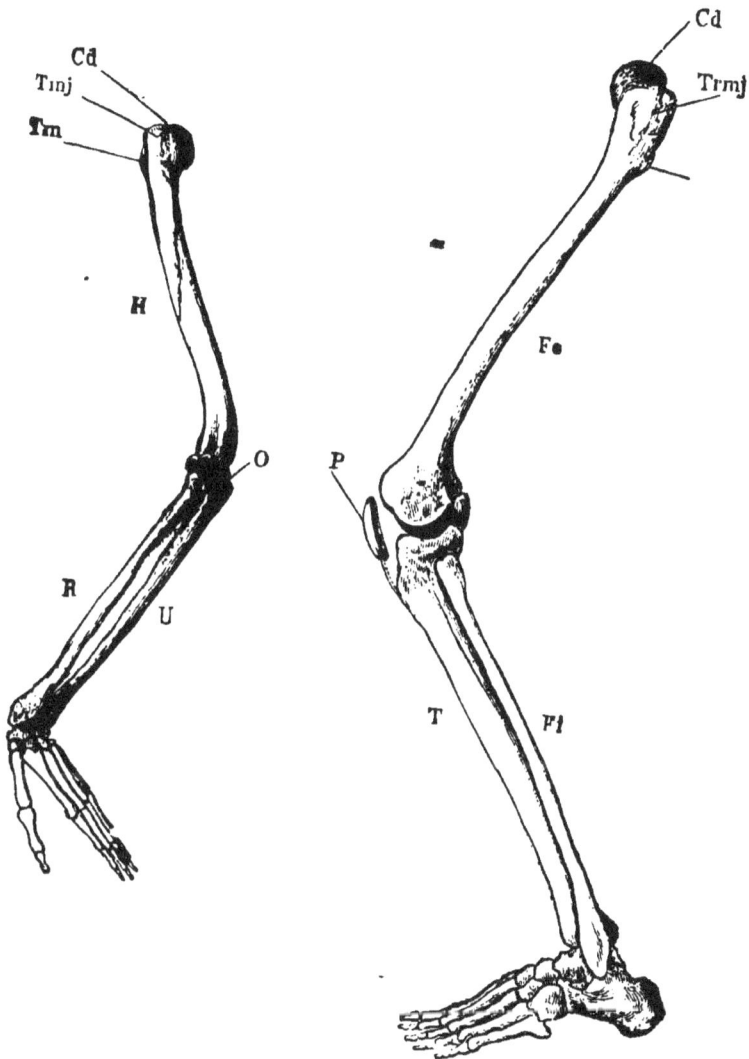

FIG. 17.—The skeleton of the arm and leg. *H,* the humerus; *Cd,* its articular head which fits into the glenoid fossa of the scapula; *U,* the ulna; *R,* the radius: *O,* the olecranon; *Fe,* the femur; *P,* the patella; *Fi,* the fibula; *T,* the tibia.

been enumerated, but certain resemblances and differences between the skeletons of the two limbs (Fig. 17) are worth noting. In general plan it is clear that they

correspond pretty closely to one another; the pectoral
arch answers to the pelvic ; the humerus to the femur ;
the radius and ulna are represented by the tibia and
fibula ; **five** metacarpal bones correspond to five metatarsal, and fourteen phalanges in the digits of the hand to
fourteen in the digits of the foot ; elbow and knee-joints,
and wrist and ankle are comparable.   There is, however,
in the arm no separate bone at the elbow answering to the
patella at the knee ; but the ulna bears above a bony
process, O, which is in early life a separate bone and
represents the patella.   There are in the adult carpus
eight bones, in the tarsus but seven ; here again we find,
however, that originally the *astragalus, Ta* (Fig. 19), of
the tarsus consists of two bones.   The elbow-joint bends
ventrally and the knee-joint dorsally.

When we compare the limbs as a whole greater differences come to light, differences which are related to their
different uses.   The arms, serving as prehensile organs,
have all their parts as movable as is consistent with the
requisite strength ; the lower limbs, having to carry all the
weight of the body, have their parts more firmly knit
together.   Accordingly we find the shoulder girdle, C, S

Do the upper and lower limbs correspond in general plan of structure ?   What in the lower limb answers to the pectoral girdle ?
What to the humerus ?   What bones to those of the forearm ?   What
to the metacarpal ?   Do the phalanges of the hand and foot agree in
number ?   What joints in the leg answer to elbow and wrist ?

What bone in the leg is not represented by a separate bone in the
arm of adults?   What in the arm corresponds to this leg bone?   Is it
ever a separate bone?   Which has more bones, hand or foot?   How
many bones are there in the tarsus in infancy?   How many afterwards unite to form one?   What is the bone formed by this union
named?   How do elbow and knee joints differ as to the direction in
which they bend?

Why are the arms made as movable as possible?   Why are the
lower limb bones more firmly knit?

Fig. 18, only directly attached to the axial skeleton by the ventral ends of the collar bones, and free to make consid-

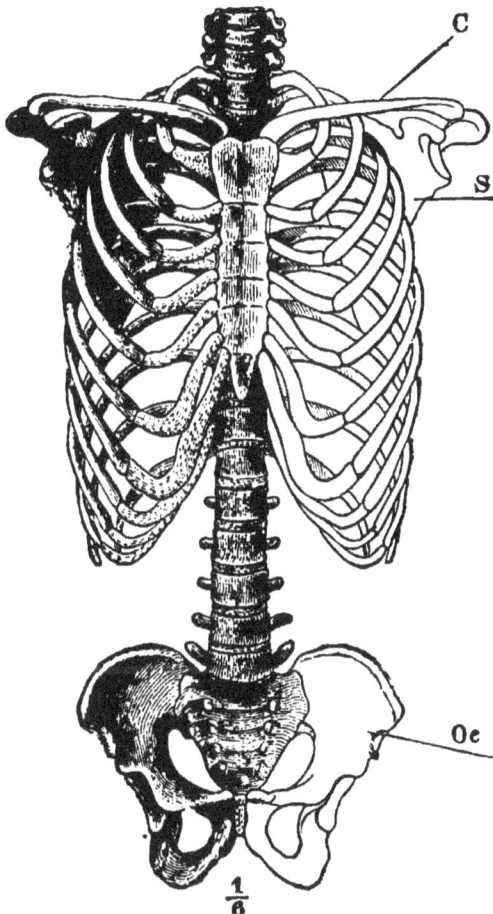

Fig. 18.—The skeleton of the trunk and the limb arches seen from the front. *C*, clavicle ; *S*, scapula ; *Oc*, innominate bone attached to the side of the sacrum dorsally and meeting its fellow at the *pubic symphysis* in the ventral median line.

erable movements, as in "shrugging the shoulders." The pelvic girdle, *Oc*, on the contrary, is firmly and immovably attached to the sides of the sacrum.

How is the shoulder girdle united to the axial skeleton? Can it move? Give an instance? How is the pelvic attached? Is it movable?

The socket on the outer end of the shoulder-blade, with which the humerus forms the shoulder-joint, is very shallow, and allows of much freer movement than is permitted by the deeper socket of the pelvis, into which the top of the thigh bone fits.

If we hold one humerus tightly and do not allow it to rotate, we can still move the forearm bones so as to turn the palm of the hand up or down; no such movement is possible between tibia and fibula.

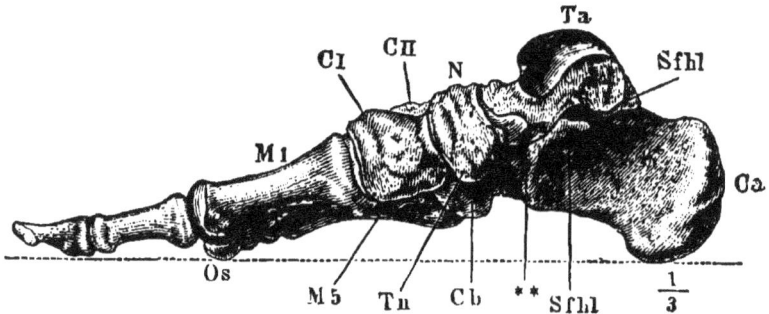

Fig. 19.—The bones of the foot. *Ca, Calcaneum,* or heel bone; *Ta,* articular surface for tibia on the *astragalus; Cb,* the cuboid bone.

In the foot the bones are much less movable than in the hand, and are so arranged as to make a springy arch (Fig. 19) which bears behind on the heel bone, *Ca,* and in front on the far ends, *Os,* of the metatarsal bones; over the crown of the arch at *Ta* is the surface with which the leg-bones articulate, and on which the weight of the body bears when we stand.

Which is deeper, the socket on shoulder-blade for humerus, or on pelvic girdle for femur?

Can the foot be turned round so as to bring its sole upwards? Can the hand so as to bring the palm up?

Are the hand or the foot bones more movable? How are the foot bones arranged? On what points does the arch of the foot bear? On what part of the arch is the weight of the body borne?

The toes are far less mobile than the fingers, the difference between great toe and thumb being especially marked. The thumb can be made to meet each of the finger-tips, and so the hand can seize and manipulate very small objects, while this power of *opposing* the great toe to the others is nearly absent in the foot of civilized man. In infants, and in savages who have never worn boots, the great toe is often much more movable, though it never acts so completely like a thumb as it does in most apes, whose feet are used for prehension nearly as much as their hands. Our own toes can by practice be made much more movable than they usually are; persons born without hands have learned to write and paint with the toes.

Peculiarities of the Human Skeleton.—There are some interesting points in the structure of the human skeleton, connected with our power of maintaining the erect posture, and of progressing on the feet so that the hands are left free for grasping. In no other vertebrate is the division of labor between the anterior and posterior limbs carried so far; the highest apes often use the hand in locomotion and the foot for prehension. As characteristic of man's skeleton we may note:

1. The skull is nearly balanced* on the top of the vertebral column (Fig. 20) so that but little effort is needed to

Are toes or fingers more mobile? How does the thumb differ in this respect from the great toe? What reason have we to think that the shoe has produced this effect? In what animals is the great toe more movable? What power have their feet in consequence? Can we make our toes more movable by practice? Illustrate.

With what facts are the more marked peculiarities of the human skeleton connected? In what living creature is the division of labor between arms and legs carried farthest?

Does the skull of man nearly balance on its support?

* The balance is, however, not quite complete. When any one goes to sleep in an ill-ventilated lecture room he is usually awakened by a sharp jerk downwards of

keep the head erect. In four-footed beasts the skull, being carried on the front end of a horizontal backbone, needs special ligaments and considerable muscular effort to support it; in apes the skull does not nearly balance on the top of the spine; its face is much heavier than its back part, while in men the face bones are relatively smaller and the cranium larger. To keep the head erect and look things straight in the face "like a man" is far more fatiguing to monkeys, and they cannot maintain that position long.

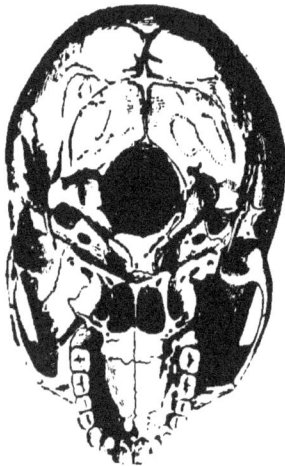

Fig. 20.—The base of the skull. The lower jaw has been removed. At the lower part of the figure is the hard palate forming the roof of the mouth and surrounded by the upper set of teeth. Above this are the paired openings of the posterior nares, and a short way above the middle of the figure is the large median *foramen magnum*, with the bony convexities (or *occipital condyles*) which articulate with the atlas, on its sides. It will be seen that the part of the skull behind the occipital condyles is about equal in size to that in front of them; in an ape the portion in front of the occipital condyles would be much larger than that behind them.

2. The human spinal column, when viewed from the front, is seen to widen gradually from the neck to the sacrum, and so to be well fitted to sustain the weight of the head, upper limbs, &c., carried by it. Its curvatures, which are peculiarly human, add greatly to its spring and elasticity; were it a straight rigid rod the brain, concealed in the skull at its top, would be jarred at every step.

How do four-footed beasts differ in this respect? Do apes' skulls balance as well as man's? Why not? What is the result of this want of balance?

What is observed when the human spinal column is viewed from the front? What is gained by its gradual widening from above down? What feature in our spines is peculiarly human? What benefit results from it?

---

his chin. The muscles concerned in holding the head erect having relaxed their vigilance the greater weight of the front half of the skull exerts its effect.

3. The pelvis, to the sides of which the lower limbs are attached, is proportionately very broad in man, so that the balance of the trunk on the legs is not easily upset when the body is bent towards one side.

4. The lower limbs are proportionately very long in man. This makes progression on them more rapid by allowing a longer stride, and also makes it difficult to go on "all fours" except by creeping on the hands and knees. The arms of some apes are as long, and of others longer, than their legs.

5. The arched instep and broad sole of the human foot are very characteristic. Most beasts, as horses, walk on the tips of their toes, the hoof being really a very big nail; others, as bears, place the heel also on the ground, but have a much less developed tarsal arch than man. The vaulted human tarsus, made up of a number of small bones, each of which can glide a little over its neighbors, but none of which can move much, is admirably calculated to break any jar which might be transmitted to the spinal column by the contact of the sole with the ground at each step.* A well arched instep is

What feature characterizes the human pelvis? What benefit results from it? Which limbs are longest in man? What ends are gained by the considerable length of the legs? Why do infants crawl on the hands and knees instead of the hands and feet? Which limbs are longest in apes?

What structural points in the foot are especially human? What part of the foot do horses put on the ground? Name an animal which puts the heel also on the ground when it walks. How does the bear's tarsal arch differ from man's?

What benefit results from the form and structure of the human tarsus? How? Why is a well-arched instep beautiful?

---

* A carriage spring consists of two curved elastic steel bars fastened together at their ends, and with their concave sides turned towards one another. The axle of the wheel is attached to the middle of the lower bar, and the weight of the carriage

therefore rightly considered beautiful; it makes the gait easier and more graceful.

---

bears on the middle of the upper. When the wheel jolts over a stone the jerk is transmitted to the elastic arches, which each flatten a little, and so instead of a sudden jerk a gentle sway is transmitted to the carriage. The tarsal arch of the human foot acts like the upper half of a carriage spring.

CHAPTER V.

THE STRUCTURE, COMPOSITION AND HYGIENE OF BONES.

The gross structure of bones.—Although the bones differ very much in shape all are alike in microscopic structure and in chemical composition. When alive they have a bluish-white color, with a pinkish hue when blood is flowing through them; they possess considerable flexibility and elasticity, which may be best observed in a long slender bone, as a rib.*

To get a general idea of the structure of a bone we may select the humerus (Fig. 21). When fresh this is closely invested on its outside by a tough membrane, *the periosteum*, composed of connective tissue and containing many blood-vessels. On its under side new bony tissue is deposited as long as the bone is growing thicker, and throughout life it is concerned in the nourishment of the bone,

How do bones differ from one another? In what respects do all bones agree? What is the color of a living bone? Name some mechanical properties of bone. In what bones may such properties be most readily seen?

What covers a bone on the exterior? What is it composed of? Does it contain bloodvessels?

---

* The rib of a sheep or a rabbit when thoroughly boiled can be readily scraped clean and preserved, and serves admirably to show the flexibility and elasticity of bone.

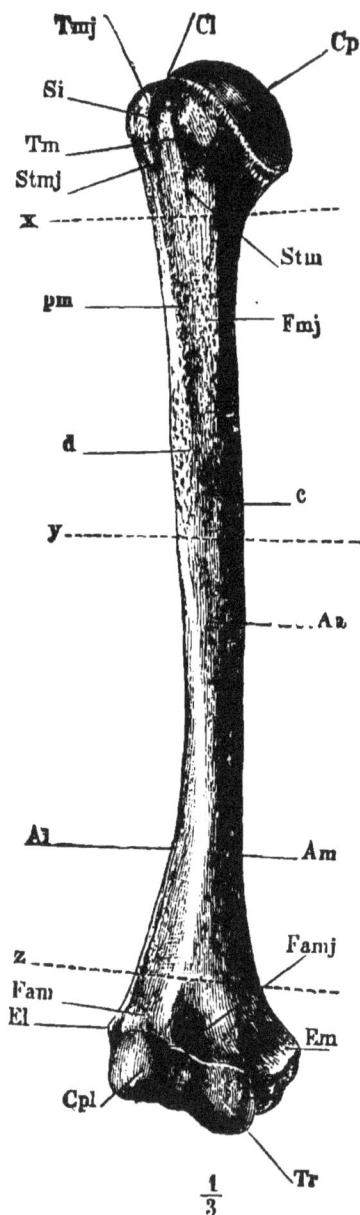

Tmj
Si
Tm
Stmj
x
pm
d
y
Cl
Cp
Stm
Fmj
c
Ae
Al
z
Fam
El
Am
Famj
Em
Cpl
Tr

$\frac{1}{3}$

Fig. 21.—The right humerus, seen from the front. For description, see text.

which dies if it be stripped off.* The periosteum covers the humerus except on its ends (*Cp*, *Tr*, *Cpl*) at the shoulder and elbow-joints; there the bone is covered by a thin layer of gristle or cartilage. Very early in life the whole humerus consists of cartilage; this is afterwards absorbed and replaced by bone, leaving only a thin layer of *articular cartilage* on each end.

The bone itself consists of a central nearly cylindrical portion or *shaft* (extending between the dotted lines X and Z) and two *articular extremities*. These extremities are enlarged to give a wider

What are the functions of the periosteum? Where is the periosteum absent? Of what does the humerus consist in very early in life? What happens to most of its cartilage afterwards? Where is some cartilage left?

What are the main divisions of the humerus? What is the general form of its shaft? Why are its articular extremities large?

* Cases have been recorded in which a considerable portion of a bone or even the whole bone has been removed during life, and the periosteum (left but slightly injured) has formed a new bone in place of the old.

bearing surface in the joints, and also to provide space on which to attach the muscles which move the bone ; the various knobs on the extremities, and the rough patches on the shaft, all mark areas where muscles were fixed.

Internal structure.—If the humerus were divided lengthwise we would find that its shaft was hollow ; the space is known as the *medullary cavity*, and in life is filled with soft fatty marrow. Fig. 22 represents such a longitudinal section. We see in it that the marrow cavity ends near the articular extremities ; and that in these the bone has a loose, spongy texture, except a thin dense layer on the surface. In the shaft the compact outer layer is much the thicker, the spongy portion only forming a thin stratum next the medullary cavity.* To the unassisted eye the spongy bone appears made up of a trellis-work of thin bony plates which intersect in all directions and surround cavities about the size of the

What do the knobs and rough patches on the bone indicate?

What should we find on dividing the humerus lengthwise? What is its shaft cavity called? What does it contain? Where does the marrow cavity end? What is the texture of the articular extremities of the bone? How does the shaft differ in structure from the extremities of the femur? What does the spongy bone look like?

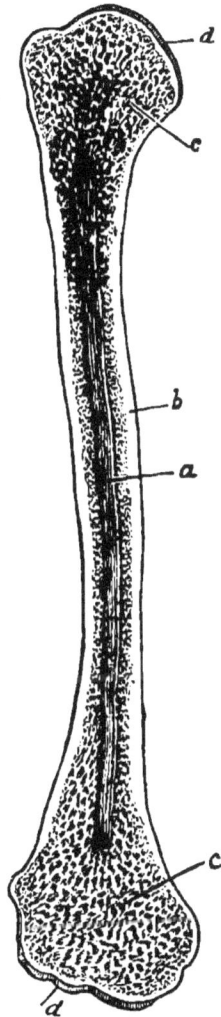

Fig. 22.—The humerus cut open. *a,* marrow cavity; *b,* hard bone; *c,* spongy bone; *d,* cartilage.

* These facts may readily be demonstrated by sawing in two lengthwise the bones out of a leg of mutton,

head of a small pin. In these spaces there is found during life a substance known as the *red marrow*, which is quite different from the yellow fatty marrow of the medullary cavity.*

Why bones are hollow.—If the bones were solid they would be extremely heavy and unnecessarily strong for the common purposes of life, unless only the same amount of material were used in their construction, and then they would be either loose in texture and easily broken, or, if dense, they would be thin rods and not give sufficient surface for the attachment of muscles! It is a well-known principle in practical mechanics that the same amount of material will bear a greater strain if in the form of a tube than in that of a solid rod of the same length; hence iron pillars are cast hollow; to fill them up solidly would make them enormously heavier without anything like a proportionate increase in strength. Take a glass tube one foot long and a piece of glass rod of the same length and weight; support each at its ends and hang weights on the middle until it breaks; the tube will be found to bear a very much greater strain before yielding. We see an application of this same method of utilizing a given amount of material to the best advantage for support, in the hollow stalks of grass, wheat, and barley.

Varieties of structure found in different bones.—Bones which, like the humerus and femur, present a shaft and

What lies in the cavities of the spongy bone?
Why are most bones hollow? Which will bear most weight, a tube or a solid rod of the same material, weight, and length? Give illustrations. Why are grass stalks hollow?

---

* Many of the bones of birds are thin-walled tubes of dense bone : the central cavity contains air and no marrow, and communicates by tubes with the lungs. Examine the humerus of a pigeon or a rooster

articular extremities, are called *long bones;* other examples are tibia and fibula, radius and ulna, metacarpal and metatarsal bones, and the phalanges of fingers and toes. *Tabular bones* form thin plates, like those of the roof of the skull, and the shoulder-blades. *Short bones* are rounded or angular, and not much longer in one diameter than another ; as the carpal and tarsal (Fig. 19) bones. *Irregular bones* include all which do not fit well into any of the above classes ; they usually lie in the middle line of the body and are divisible into similar right and left halves ; the vertebræ are good examples.

All bones are covered by periosteum except where they enter into the formation of a joint, but in the human body only the long bones possess a medullary cavity containing yellow marrow. The rest are filled up by spongy bone, covered by a thin layer of dense, and have red marrow in their spaces.

The histology of bone.—The microscope shows that compact bone is only so to the naked eye ; even a hand lens shows minute holes in it; it but differs from spongy bone in the fact that its cavities are much smaller, and the hard bony plates between them thicker. If a thin transverse section of the shaft of long bone (Fig. 23) be examined with a microscope magnifying about twenty diameters, even its densest part will be seen to show numerous openings which become gradually larger near the medullary

What is a long bone? Give examples. A tabular bone? Examples. A short bone? Examples. An irregular bone? Examples.

With what is most of the surface of bones covered? Where is the periosteum absent? What bones contain yellow marrow? What do the others contain?

Does compact bone contain any cavities? How may these be seen ? How does it differ from spongy bone ? What is seen when a thin slice of bone is magnified twenty times? Where do the apertures in it become larger?

cavity and pass insensibly into the spaces of the spongy
bone around it. These openings are the cross sections of

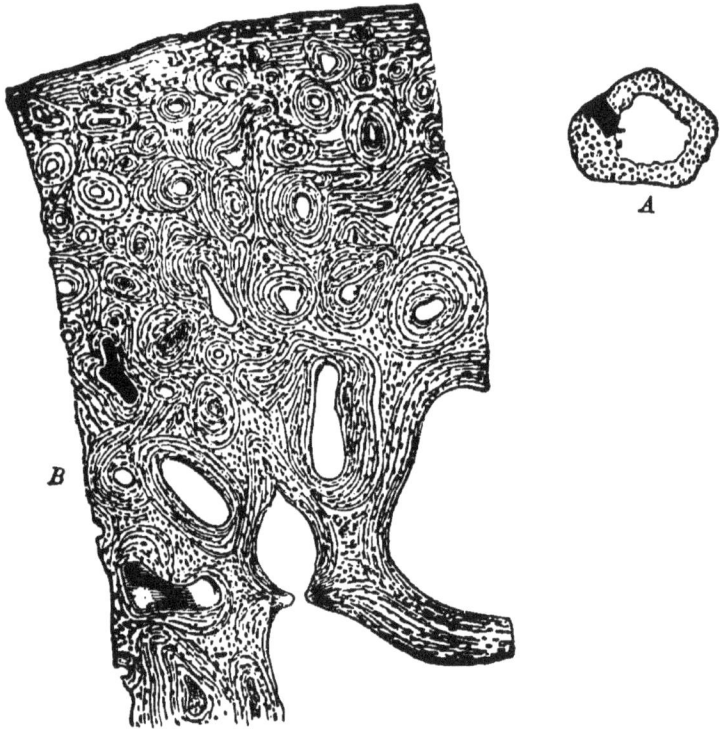

Fig. 23.—*A*, a transverse section of the ulna, natural size; showing the medullary cavity. *B*, the more deeply shaded part of *A* magnified twenty diameters.

tubes known as the *Haversian canals,* which run all
through the bone, the majority of them in the direction of
its long axis, though numerous cross branches unite them.
The outermost Haversian canals open on the surface of
the bone beneath the periosteum; from there blood-vessels pass in, and, traversing the whole bone in these channels, convey materials for its growth and nourishment.

What are the Haversian canals? Where do the outer ones open?
What enters them? Where do the blood-vessels of a bone run?
What do they carry?

Around each Haversian canal is a series of plates or *lamellæ*, each canal and its lamellæ forming an *Haversian system ;* the entire bone is made up of a number of such systems, with the addition of a few lamellæ lying in the corners between them, and some which run around the whole bone on its outer surface. In the spongy bone the Haversian canals are very large, containing red marrow as well as blood-vessels, and the lamellæ around each are few in number.

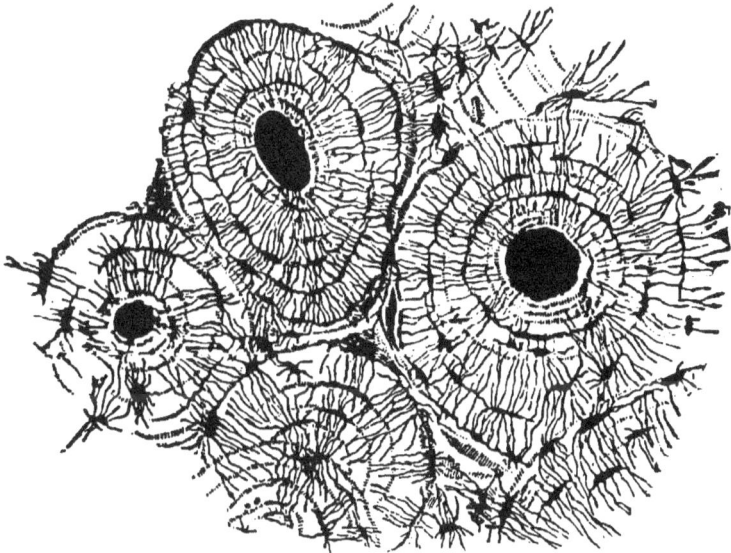

Fig. 24.—A small piece of bone, ground very thin and highly magnified.

If a bit of bone be still more magnified (Fig. 24) we find that very small cavities lie between the lamellæ; they

What lies around each Haversian canal? What is an Haversian system? Of what does a bone consist in addition to Haversian systems? Are the Haversian canals comparatively large or small in spongy bone? What do the spaces of spongy bone contain in addition to blood-vessels?
What spaces lie between the lamellæ of an Haversian system?

are called *lacunæ;* from each lacuna radiate many extremely fine tubes, the *canaliculi,* so that each lacuna with its canaliculi looks something like a small animal with a great many legs. The innermost canaliculi open into the Haversian canal of the system to which they belong, and those of various lacunæ communicate with one another, so that a set of passages is provided through which liquid which transudes from the blood-vessel in the Haversian canal can ooze all through the bone.

In a living bone a nucleated cell lies in each lacuna. These cells, *the bone corpuscles,* are the remnants of those which made the bone, all whose hard parts are but intercellular substance: a sort of skeleton is made by each cell around itself, and this adheres to the skeletons of the rest, and thus the whole bone is built.

Chemical composition of bone.—Apart from the bone corpuscles and the soft contents of the Haversian canals and of the spaces of the cancellated bone, the hard bony substance proper is composed of animal and mineral matters so intimately combined that the smallest distinguishable bit of bone contains both. The mineral matters give the bone its hardness and stiffness, and form about two-thirds of its weight when dried. They may be removed by soaking the bone in diluted muriatic acid,* and the

What radiate from the lacunæ? Into what do the innermost canaliculi of an Haversian system open? How is nourishing liquid from the blood carried throughout a bone?

What lies in each lacuna of a living bone? What are the bone corpuscles? What is the hard part of bone? How is a whole bone made up?

Of what primary constituents is bone composed? Is there any fragment of bone that does not contain both? What qualities do its mineral parts give to a bone? How much of a dry bone consists of mineral matter? How may the mineral parts be removed?

* Add a couple of ounces of muriatic acid to a pint of water and place a

animal or organic part of the bone is then left as a tough, flexible mass, retaining perfectly the shape of the original bone.

When long boiled in water the greater part of the animal portion of bone is turned into *gelatine* and dissolved in the water; most of the gelatine which we buy in the shops is obtained by boiling fresh bones in a closed vessel under a high pressure; the water then gets much hotter than when boiled in the air, and dissolves out the gelatine more quickly; when a shin of beef is used to make soup the bones are put in as well as the softer parts, and the whole is kept boiling for hours so as to get some of the gelatine out of the bones. The animal matter of bone gives it its toughness and flexibility.

The earthy portion may be obtained free from the . animal by calcining a bone in a bright fire. The residue is a white and very brittle mass, which retains perfectly the shape of the original bone. It is readily powdered and then forms bone ash, which consists chiefly of phosphate and carbonate of calcium; most of the phosphorus of commerce is obtained from it. If the burning be imperfect the animal matter is charred but not altogether burnt away, and a black mass, known as animal charcoal or "bone black," is left.

What then remains behind? What are its properties? Has it still the shape of the bone? What happens when a bone is boiled for hours? How is the gelatine of commerce obtained? Why do we use bones in making soup? What properties does its animal matter confer on bones?

How may we get the mineral part of bone free from the animal? What are its properties when isolated? What is bone ash? From what is phosphorus prepared? What is animal charcoal?

sheep's rib in the mixture for four or five days, having previously scraped the bone quite clean. It will be found so flexible that a knot may be tied on it; the specimen may be preserved in strong brine or dilute alcohol from year to year for exhibition to a class.

Hygiene of the bony skeleton.—In early life the animal matter of the bones is present in larger proportion than later; hence the bones of children are tougher, more pliable, and not so easily broken. The bones of a young child are tolerabiy flexible and are capable of being distorted by any long-continued strain ; therefore children should never be placed on a bench so high that the feet have no support ; if this is frequently done the thigh bones will almost certainly become bent over the edge of the seat by the weight of the lower legs and the feet, and a permanent distortion may be produced. For the same reason it is important that a child be made to sit straight, when writing or drawing, to avoid the risk of producing a lateral curvature of the spinal column ; and young children should not be made to walk too early lest they become bow-legged, their bones not being rigid enough to bear the weight of the body. How easily the bones yield to prolonged pressure in early life is well illustrated by the distorted feet of Chinese ladies ; and by the extraordinary forms (Fig. 25) which some races produce in their skulls by tying boards or bandages on the heads of the children. A distorted foot, even in the United States, is no uncommon thing in these days of tight boots and high heels. The latter are especially bad, as, instead of allowing the weight of the body to bear properly directly downwards on the crown of the arch of the instep, they throw it forwards, and violently force the fore part of the foot into the toe of

How do the bones of children differ in composition from those of adults? How in properties? Why should children have a seat allowing the feet to be supported? Why should young persons be taught to sit erect while writing? What will happen if young children are encouraged to walk too much? Why? Give instances of the readiness with which bones can be distorted in early life. Why are high-heeled boots hurtful?

the boot. This not only crushes the toes and leads to deformities, corns, and bunions, but makes the gait stiff, inelastic and ungraceful.

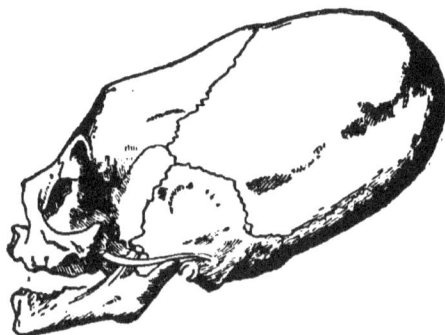

Fig. 25.—Skull of a child of the tribe of Chinook Indians (inhabiting the neighborhood of the Columbia river), distorted by tight bandaging so as to assume the shape considered elegant and fashionable by the tribe.

In advanced life the animal matter of the bones is present in deficient amount, and hence they are brittle and easily broken.

An infant has its bones but very imperfectly hardened by mineral matter. Hence the great importance of supplying it with food containing phosphate of lime, which is the chief mineral constituent of bone. Of all common articles of diet, milk contains most phosphate of lime : hence one great reason of its value as a food for children.

Fracture.—When a bone is broken it is said to be *fractured;* when it is a clean break the fracture is *simple;* when the bone is more or less broken up into bits on each side of the break the fracture is *comminuted;* when the

Why are the bones of an old person easily broken? Why should milk form part of a child's diet ?

What is a fracture? A simple fracture? A comminuted fracture? A compound fracture?

soft parts also are lacerated, so that there is an opening from the skin to the broken bone, the fracture is *compound.*

Once a bone is broken the muscles attached to it are apt to pull its ends out of place; hence it requires to be "set," and then kept in position by splints or bandages; this frequently needs much skill and a thorough knowledge of the anatomy of the body. A medical man should be summoned at once, as the parts around the break commonly swell very rapidly and make the exact nature of the fracture hard to detect, and also the replacement of the displaced ends more difficult.

It is known to physicians that serious bone diseases are sometimes due to the continued drinking of alcohol; also that the action of alcohol on the periosteum may render that membrane incapable of forming healthy bone from materials supplied by the blood; thus causing bad nutrition, leading to disease of the bones. Surgeons find that broken bones of habitual drinkers do not unite as readily as those of other patients.

Why does a broken bone need "setting"? What is the object of "splints"? Why should skilled assistance be obtained as soon as possible after a bone has been fractured?

In what way does alcohol tend to promote bone diseases? To interfere with the repair of fractures?

Appendix to Chapter V.

When giving lessons on Chapters IV and V, it is very desirable for a teacher to have at hand an articulated human skeleton. This may be purchased for about $40.00 from Henry Ward, Rochester, N. Y., and will last for an indefinite number of years. When the school funds do not permit the purchase of a skeleton, one can almost certainly be borrowed from some medical man or medical school for a few days. When there are several public schools in a city it would probably be possible to induce the school commissioners to purchase a skeleton to be used by the schools in turn.

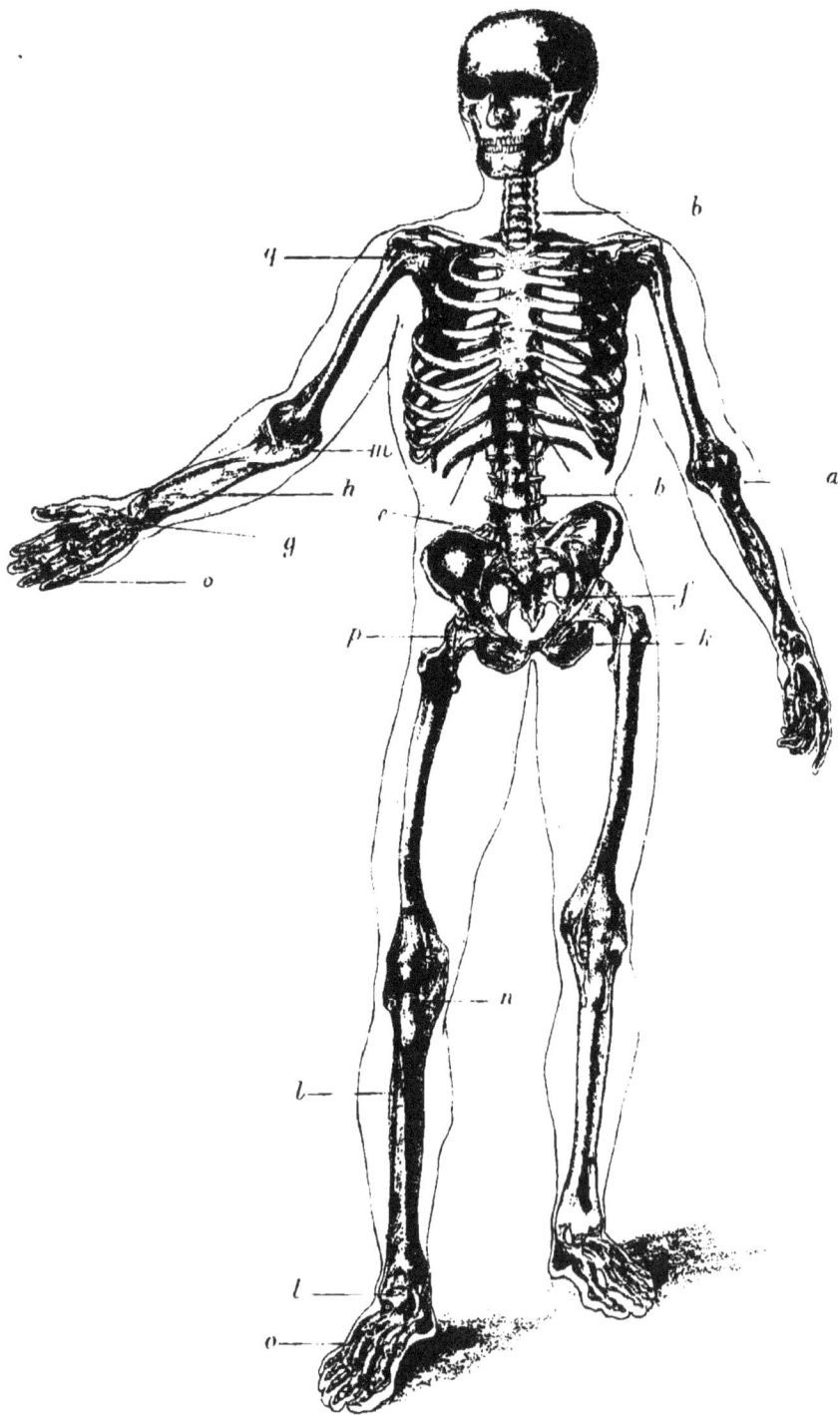

PLATE I.—THE BONES, JOINTS, AND LIGAMENTS.

EXPLANATION OF PLATE I.

A front view of an adult human skeleton to illustrate the mode in which the bones are connected together at the different joints.

For the names of the bones consult the description of figure 8, which commences on page 38.

a Ligaments of the Elbow-Joint.
b The Ligament which is connected to the ventral surfaces of the bodies of the Vertebræ.
e Ligament connecting the innominate Bone to the Spine.
f Ligament connecting the innominate Bone to the Sacrum.
g The Ligaments of the Wrist Joint.
h The Membrane which fills up the interval between the two bones of the Fore Arm.
l A similar Membrane between the two bones of the Leg, and, lower down, *l*, ligaments of the Ankle-Joint.
k A Membrane which fills up a hole in the Innominate Bone.
n Ligaments of the Knee-Joint.
o o Ligaments of the Toes and Fingers.
p Capsular (bag-like) Ligament of the Hip-Joint.
q Capsular Ligament of the Shoulder-Joint.

CHAPTER VI.

JOINTS.

The movements of the body are brought about by means of soft reddish organs known as *the muscles;* the lean of meat is muscle, so every one knows what a dead muscle looks like.* Muscles have the power of shortening with considerable force ; when they do so they pull their ends towards one another and swell out in the middle ; in other words, they become shorter and thicker. With few exceptions the ends of a muscle are attached to separate bones † between which a joint lies, and when the muscle shortens, or, in physiological language, *contracts*, it produces movement at the joint. The joints and muscles thus form the chief motor apparatuses of the body.

What organs produce the movements of the body? What is the technical name of the lean of meat? What power do muscles possess? What happens when they exert it? To what are the ends of most muscles attached? What happens when the muscle contracts? Name the chief motor apparatuses of the body?

---

* In many animals some muscles are much redder than others, and it is then found that the deeper colored are those which are kept most constantly in use ; the leg muscles of a chicken, for example, are redder ,than those of the wings and breast, and as the coloring matter is turned brown by heat, they form the "dark meat" after cooking ; in birds which fly a great deal the breast muscles (which chiefly move the wings) are also dark. The heart, which is a muscle always at work, is deep red, even in fishes, most of whose muscles are pale.

† As an example of a muscle not attached to the skeleton, we may take the *orbicularis oris*, which forms a ring around the mouth-opening beneath the skin of the lips : when it contracts it closes the mouth, or if it contracts more forcibly purses out the lips. The *orbicularis palpebrarum* forms a similar ring around the eye opening, and when it contracts closes the eye.

Joints.—Articulations which permit of movement by the gliding of one bone over another are called *joints;* all are constructed on the same general plan, though the range and direction of movement permitted are different in different joints. As an example we may take the hip-joint, a section through which is represented in Fig. 26.

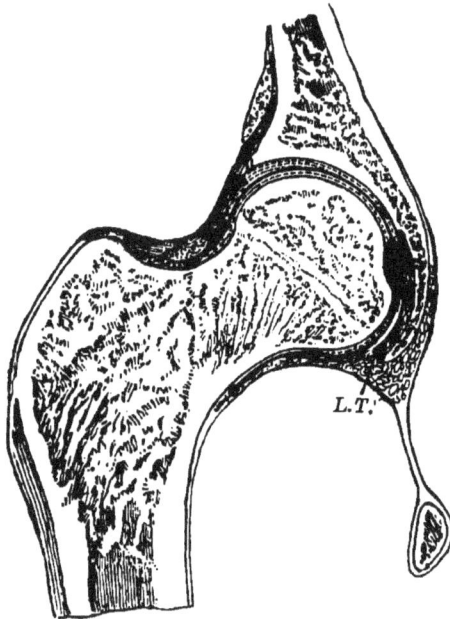

FIG. 26.—Section through the hip-joint.

On the outer side of the os innominatum (*s*, Fig. 8) is a deep hollow, *the acetabulum*, which receives the upper end of the thigh-bone. The acetabulum is lined by a thin layer of cartilage, with an extremely smooth surface, and its cavity is also deepened by a cartilaginous rim. The upper end of the femur consists of a nearly spherical head, borne on a narrower neck ; this head is covered by cartilage, and rolls smoothly in the acetabulum like a ball in

What is a joint? How do joints differ? Describe the hip-joint.

a socket. If the hard bones came into direct contact they would be apt to chip one another when a sudden movement was made, especially if the hip-joint were so far bent as to knock the thigh-bone against the rim of the acetabulum; the elastic and yielding cartilage forms a protecting cushion between the bones and prevents this.

To keep the bones in place and limit the range of movement, *ligaments* pass from one to the other ; they are composed of connective tissue, are extremely pliable but cannot be stretched, and are very tough and strong.   One is the *capsular ligament,* which forms a bag all round the joint, and another is the *round ligament, L. T.,* Fig. 26, which passes from the rim of the acetabulum to the head of the femur ; from the rim of the socket it passes to the center of the acetabulum along a groove in the bone, and then turns out to be fixed to the thigh-bone.

Covering the inside of the capsular ligament and continued over the cartilages of the joint is the *synoviai membrane,* very thin and composed of a layer of flat cells. This pours out into the joint a very small quantity of *synovial liquid,* which is somewhat like the white of a raw egg in consistency, and plays the part of the oil moistening those surfaces of a machine which glide over one another ; it lubricates the joint and enables all to run smoothly and with but little friction.

In the natural state of the parts the synovial membrane

What is the use of the cartilage lining the bones which move over one another in a joint?
What is the use of ligaments? Of what are they composed? What are their properties? Name some ligaments of the hip-joint. Where does the capsular ligament lie? Where the round ligament? What membrane lines the joint? Of what is it composed? What does it pour into the joint? What is synovial liquid like? What is its use? Illustrate by an example.

on the head of the thigh-bone lies close against that lining the acetabulum, so that practically there is no cavity left in the joint. This close contact is not maintained by the ligaments (which are much too loose, and serve mainly to prevent such excessive movement as might roll the femur quite out of its socket), but by the many strong muscles which pass between pelvis and thigh-bone and hold both firmly together. In addition, the pressure of the atmosphere is transmitted by the skin and muscles to the exterior of the air-tight joint, and helps to keep its surfaces together. If all the muscles be cut away from around the hip-joint of a dead body, it is found that the head of the femur is still held in its place by the pressure of the air ; and so firmly that the weight of the whole limb will not draw it out ; but if a hole be pierced into the bottom of the acetabulum, and air be thus let into the joint, then the thigh-bone falls out of place as far as the ligaments will let it.

In all joints we find the same essential parts; bones, articular cartilages, synovial membrane, synovial liquid, and ligaments.*

Ball and socket joints.—Such a joint as that at the hip

Is there in health any definite space between the bones of the hip-joint? What is the chief use of the ligaments? How are the bones held together? What in addition to muscles helps to keep the bones of the joint in contact? Describe an experiment illustrating the effect of atmospheric pressure in keeping the bones together?

What essential parts are found in all joints?

What is such a joint as the hip-joint called?

---

* The structure of joints can be readily seen in those of a fresh calf's or sheep's foot. The synovial membrane is so thin and so closely adherent to the parts it lines that a microscope is needed for its demonstration; but all the other parts are readily made out.

is called a ball and socket joint, and allows of a greater variety of movement than any other kind. Through movements taking place at it the thigh can (1) be *flexed,* that is, bent so that the knee approaches the chest, and (2) *extended* or straightened again ; it can (3) be *abducted* so that the knee is moved away from the middle line of the body, and (4) *adducted* or brought back again ; by movement at the hip the limb can also (5) be *circumducted,* so that, with knee and ankle joints held rigid, the whole leg is made to describe a cone, of which the apex is at the hip-joint and the base at the foot ; and finally (6) *rotated* so that the whole limb can be rolled to and fro a little about its own long axis. All ball and sockets joints allow all these movements to a greater or less extent.

Another important ball and socket joint is that between the upper end of the humerus and the hollow (*glenoid fossa*) near the upper outer corner of the shoulder blade. The glenoid fossa being much shallower than the acetabulum the range of movement possible at the shoulder, is greater than at the hip-joint.

Hinge-joints.—In this form the bony cavities and projections are not spherical, but are grooved and ridged so that one bone can glide over the other in one plane only, to and fro, like a door on its hinges.

The knee is a hinge-joint ; it can only be bent and straightened, in technical language, *flexed* and *extended.*

What kind of joints allow of the freest movement? What is meant by flexion of the thigh? By extension? By abduction? By adduction? By circumduction? By rotation? What movements do all ball and socket joints permit?

What sort of a joint is that at the shoulder? Why is more movement possible at it than at the hip-joint?

What is a hinge-joint? Give an illustration.

Name other hinge-joints.

Between the phalanges of the fingers we find also hinge-joints ; another is found between the lower jaw and the cranium, allowing us to open and close the mouth. The latter is not, however, a perfect hinge-joint ; it permits also of slight lateral movements, and a gliding motion by which the lower jaw can be thrust forward so as to bring the lower range of teeth outside the upper.*

Pivot-joints.—In this form one bone rotates about another. A good example is found between the first and second cervical vertebræ (Figs. 13, 14). The odontoid process of the axis reaches up into the neural arch of the atlas, and, kept in place there by the transverse ligament which does not let it press against the spinal cord, forms a pivot around which the atlas rotates, carrying the skull with it when we turn the head to right or left.

A more complicated kind of pivot-joint is found in the forearm. Lay the forearm and hand flat on a table, palm uppermost ; without moving the shoulder-joint at all the hand can then be turned over so that its back is upward. In this movement the radius, which carries the hand, crosses over the ulna. When the palm is turned up (*supination*) the radius and ulna are parallel (Fig. 27, A), and the radius on the outside ; place a finger of the other

Is the joint of the lower jaw with the skull a perfect hinge joint? What movements can take place at it?

What is a pivot-joint? Name an instance from the spinal column. Describe the joint between atlas and axis. What happens to the head when the atlas rotates on the odontoid process of the axis?

Where do we find another kind of pivot-joint? Illustrate its action. What happens when we turn the hand so that the palm instead of being up shall be down? How can we observe the relative change in position of radius and ulna while making this movement?

---

* The object of these minor movements is to allow us to *chew* our food ; in carnivora, as cats, which bite, but do not chew, the lower jaw forms a perfect hinge-joint with the cranium.

hand on it near the wrist, and then turn the hand over; the lower end of the radius will be found to cross over the ulna and to be on its inner side (Fig. 27, B), when the movement is completed; in this position the hand is said to be in *pronation.*

The lower end of the humerus (Fig. 21) has a large articular surface; on the inner two-thirds of this, *Tr*, the ulna fits, and the grooves and ridges of the bones interlocking form a hinge-joint, allowing us only to bend or straighten the elbow-joint. The radius fits on the rounded outer third, *Cpl*, and rotates there when the hand is turned over, the ulna forming a fixed bar around which it moves.

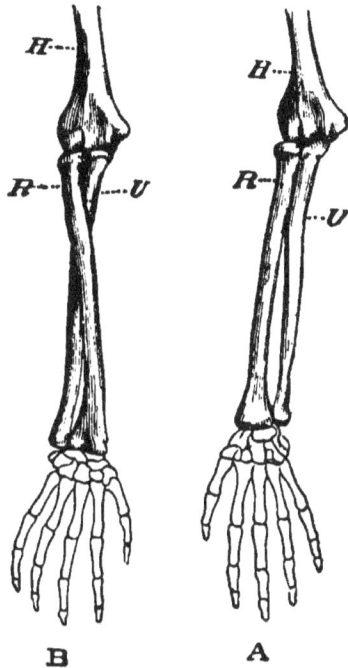

Gliding joints as a rule permit of but little movement.

Fig. 27.—A, arm in supination; B, arm in pronation; *H*, humerus; *R*, radius; *U* ulna.

Examples are found between the closely-packed bones of the carpus and tarsus (Fig. 19), which slide a little over one another when subjected to pressure.

Dislocations.—When a bone is displaced at a joint or *dislocated*, the ligaments are more or less torn and other

What movement is allowed between ulna and humerus? What between radius and humerus? Around what does the radius rotate when we turn the hand over?

Do gliding joints allow free movement? Give instances of gliding joints.

What is a dislocation? What parts are injured when a joint is dislocated?

surrounding soft parts injured. This generally leads to inflammation and swelling, which make it difficult to find out in what direction the bone has been displaced, and also greatly add to the difficulty of replacing it, or, in surgical language, of *reducing the dislocation.* The muscles attached to it are, moreover, apt to pull the dislocated bone more and more out of place. Medical aid should therefore be obtained as soon as possible ; in most cases the reduction of a dislocation can only be attempted with safety by one who knows the forms of the bones and possesses sufficient anatomical knowledge to recognize the direction of the displacement.*

A **sprain** is an injury to a joint, accompanied by straining, twisting, or tearing of-the ligaments, but without dislocation of the bones. A sprained joint should get immediate and complete rest, continued for weeks if necessary ; if there be much swelling or continued pain, medical advice should be obtained. Perhaps a greater number of permanent injuries result from neglected sprains than from broken bones. It has been found that a moderate sprain, which in a healthy adult would readily get well, in a person of alcoholic habits often results in an inflammation of the membranes in and about the joint which it is both hard and tedious to cure.

What results from this injury? What is meant by "reducing a dislocation"? Why should medical aid be obtained as soon as possible after a joint has been dislocated?
What is a sprain? How should a sprained joint be treated? What should be done at once if there is much swelling or continued pain? Are neglected sprains apt to lead to permanent injury? What has been observed by medical men' concerning sprains in persons of alcoholic habits? What is stated concerning gout?

---

* Dislocations of the fingers can usually be reduced by strong pulling, aided by a little pressure on the parts of the bones nearest the joint. The reduction of a dislocation of the thumb is much more difficult, and can rarely be accomplished without skilled assistance,

Gout is pre-eminently an alcoholic disease. An authority says: "It is *always* the effect of alcohol, either in the individual or his ancestors." This is perhaps putting the matter a little too strongly: lead-poisoning (as in the case of house-painters), or too much albuminous food in the diet, combined with a physically idle life, may lead to gout; but in the majority of cases alcohol is the cause.

EXPLANATION OF PLATE II.

A view of the muscles situated on the front surface of the body, seen in their natural position. It must be understood that beneath these muscles many others are situated, which cannot be represented in the figure.

Muscles of the Face, Head, and Neck:

1. Muscle of the Forehead. This, together with a muscle at the back of the head, has the power of moving the scalp.
2. Muscle that closes the Eyelids. The muscle that raises the upper eyelid so as to open the eye, is situated within the orbit, and consequently cannot be seen in this figure.
3, 4, 5. Muscles that raise the Upper Lip and angle of the Mouth.
6, 7. Muscles that depress the Lower Lip and angle of the Mouth. By the action of the muscles which raise the upper lip, and those that depress the lower lip, the lips are separated.
8. Muscle that draws the Lips together.
9. Muscle of the Temple (Temporal Muscle).
10. Masseter Muscle. 9 and 10 are the two chief muscles of mastication, for when they contract, the movable lower jaw is elevated, so as to crush the food between the teeth in the upper and lower jaws.
11. Muscle that compresses the Nostril. Close to its outer side is a small muscle that dilates the nostril.
12. Muscle that wrinkles the Skin of the Neck, and assists in depressing the lower jaw.
13. Muscle that assists in steadying the Head, and also in moving it from side to side.
14. Muscles that depress the Windpipe and Organ of Voice. The muscles that elevate the same parts are placed beneath the lower jaw, and cannot be seen in the figure.

Muscles that connect the upper extremity to the trunk. Portions of four of these muscles are represented in the figure, viz.:

15. Muscle that elevates the Shoulder. Trapezius Muscle.
17. Great Muscle of the Chest, which draws the Arm in front of the Chest (Great Pectoral Muscle).
18. Broad Muscle of the Back, which draws the Arm downwards across the back of the Body (Latissimus Dorsi).
19. Serrated Muscle extends between the Ribs and Shoulder-blade, and draws the shoulder forwards and rotates it, a movement which takes place in the elevation of the arm above the head (Serratus magnus).

At the lower part of the trunk, on each side, may be seen the large muscle which, from the oblique direction of its fibres, is called,

20. Outer Oblique Muscle of the Abdomen.

Several muscles lie beneath it. The outline of one of these,

21. Straight Muscle of the Abdomen, may be seen beneath the expanded tendon of insertion of the oblique muscle. These abdominal muscles, by their contraction, possess the power of compressing the contents of the abdomen.

Muscles of the upper extremity.

16. Muscle that elevates the Arm (Deltoid Muscle).
22. Biceps or Two-headed Muscle (see also page 88).
23. Anterior Muscle of the Arm. This and the Biceps are for the purpose of bending the Fore-Arm
24. Triceps, or Three-headed Muscle. This counteracts the last two muscles, for it extends the Fore-arm.
25. Muscles that bend the Wrist and Fingers, and pronate the Fore-arm and Hand—that is, turn the Hand with the palm downwards. They are called the Flexor and Pronator Muscles.
26. Muscles that extend the Wrist and Fingers, and supinate the Fore-arm and Hand—that is, turn the Hand with its palm upwards. They are called the Extensor and Supinator Muscles.
27. Muscles that constitute the ball of the Thumb. They move it in different directions.
28. Muscles that move the Little Finger.

Muscles which connect the lower extremity to the pelvic bone. Several are represented in the figure.

29. Muscle usually stated to have the power of crossing one Leg over the other, hence called the Tailor's Muscle, or Sartorius; its real action is to assist in bending the knee.
30. Muscles that draw the Thighs together (Adductor Muscles).
31. Muscles that extend or straighten the Leg (Extensor Muscles). The muscles that bend the leg are placed on the back of the thigh, so that they cannot be seen in the figure.

Muscles of the leg and foot:

32. Muscles that bend the Foot upon the Leg, and extend the Toes.
33. Muscles that raise the Heel—these form the prominence of the calf of the Leg.
34. Muscles that turn the Foot outwards.
35. A band of membrane which retains in position the tendons which pass from the leg to the foot.
36. A short muscle which extends the Toes.

The muscles which turn the foot inwards, so as to counteract the last named muscles, lie beneath the great muscles of the calf, which consequently conceal them. The foot possesses numerous muscles, which act upon the toes, so as to move them about in various directions. These are principally placed on the sole of the foot, so that they cannot be seen in the figure. Only one muscle, 36, which assists in extending the toes, is placed on the back of the foot.

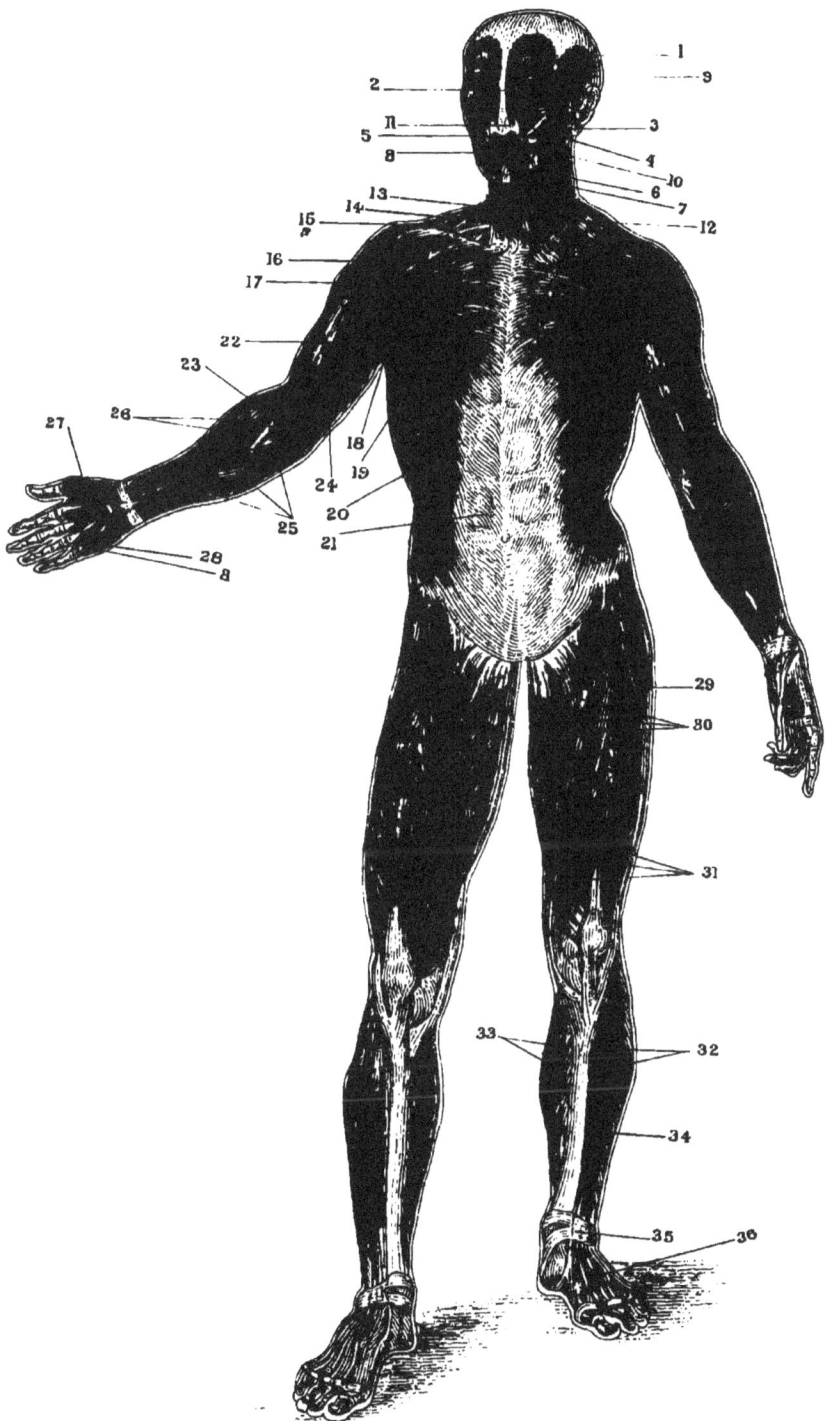

PLATE II.—THE SUPERFICIAL MUSCLES OF THE FRONT OF THE BODY.

CHAPTER VII.

THE MUSCLES.

The muscles of the human body are more than five hundred in number; they vary very much in size; from tiny ones not an inch long, in the voice-box, to that on the front of the thigh (29, Pl. II.), which passes from the pelvis to the tibia, and is eighteen inches or more in length. Whatever their size, muscles present a similar structure and possess the same properties, their various uses depending on the different directions in which they pull, and the different things they pull upon. In addition to their primary function of moving the body the muscles give it roundness and shapeliness; they also help to enclose cavities, as the abdomen and the mouth; and they hold bones together at joints.

The parts of a muscle.—In its commonest form a muscle consists of a red soft central part, called its *belly*, which tapers towards each end and there passes into one or more dense white cords, made of connective tissue and called *tendons;* the tendons attach the muscle to parts of the

About how many muscles are there in the body? Between what limits do they differ in size? In what respects do all muscles resemble one another? How are their different uses determined? What functions do muscles fulfill besides moving the body and its parts? Give examples.

What is the most usual structure of a muscle? What is the use of tendons ?

skeleton.* In Fig. 28 are shown some of the muscles of the arm. Their anatomical names we need not trouble about; but it will be seen that some (8, 11, 12) pass from arm to forearm: others, as 16, 15, 14, 13, 17, 18, start from the forearm bones and pass to the bones of the hands; near the wrist they end in slender tendons, which are bound down into place by a stout cross band of connective tissue. The skin has been dissected away from the back of the middle finger to show the endings of tendons on its phalanges.

The belly of a muscle is its

What portion of a muscle is its working part?

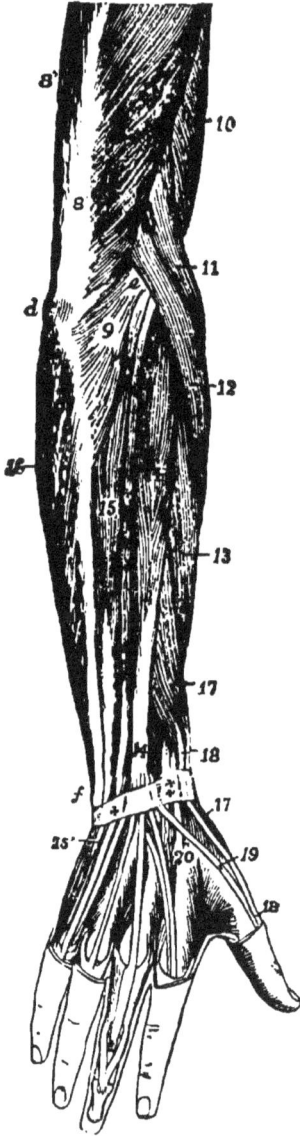

Fig. 28.—The muscles on the back of the hand, forearm, and lower half of the arm, as exposed on dissecting away the skin.

* The parts of a muscle may readily be seen in that which forms the calf of a frog's leg. Put a teaspoonful of ether in a quart of water, immerse a frog in it, and cover the vessel. In a minute the animal will be quite insensible; its head can then be cut off and its spinal cord destroyed by running a pin along it, without causing the animal any pain. Now make circular cuts through the skin at the top of the thighs and then peel the skin off like a pair of hose: it will come quite easily except about the knee-joint, where it may be necessary to carefully divide one or two tough bands. On the skinned leg many muscles will be observed, and the long slender tendons which run to the toes. The calf muscle will be seen to end below in a strong tendon near the heel. If this be divided, and the muscle turned upwards, it will be found to have at the upper end of its thick rounded belly a pair of short tendons.

working part; nerves from the brain or spinal cord enter it, and when its nerve is excited, whether involuntarily or by what we call an act of "will," the belly *contracts;* it forcibly changes its shape so as to become shorter and thicker. In so doing it drags on the tendons, which are passive inextensible cords and transmit the pull to the parts to which they are attached.

The tendons are often quite long, as for example those of many of the muscles moving the fingers (Fig. 28), whose bellies are in the forearm. The belly of the common extensor muscle of the fingers (14, Fig. 28) is seen, for example, to be in the upper half of the forearm, and to end above the wrist in a single tendon which divides up into strips which run along the back of each finger; the muscles which straighten the thumb, 17, 18 and 19, are also seen to have long slender tendons.

Where a muscle passes over a joint it is usually reduced to a narrow tendon; the bulky bellies, if they lay there, would make the joints clumsy and limit their mobility. Some muscles pass over two joints and can produce movement at either; the *biceps* of the arm, fixed above to the scapula and below to the radius, can produce movement at either the elbow or the shoulder joint.

The shortening of a muscle when it contracts is shown by the movement which it causes; the thickening may be

What enters the belly of a muscle? How is the nerve of a muscle excited? What happens when the nerve is excited? What results from the contraction of the belly of a muscle?

Are tendons ever long? Describe the common extensor muscle of the fingers and its tendons. Describe the position and length of tendons of the muscles which extend the thumb.

What happens to a muscle when it passes over a joint? Why? Name a muscle which crosses two joints. At what joints can the biceps muscle of the arm produce movement?

How is the shortening of a contracting muscle shown?

seen and felt on the biceps in front of the humerus when the elbow is bent, or in the ball of the thumb when that digit is moved so as to touch the little finger ; when a muscle contracts its belly may also be felt to grow harder. The swelling and hardening of a contracted muscle are daily illustrated when one schoolboy invites another to feel his " biceps."

The Origin and Insertion of Muscles.—Almost invariably that part of the skeleton to which one end of a mus-

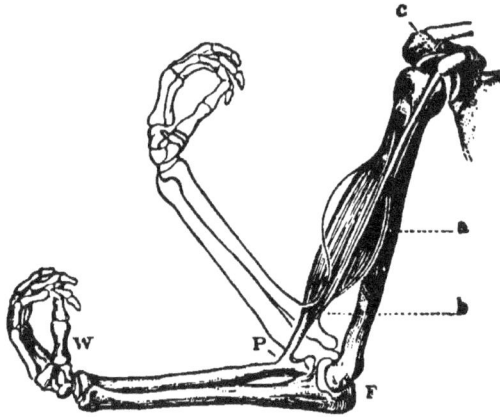

FIG. 29.—The biceps muscle and the arm-bones, to illustrate how, under ordinary circumstances, the elbow joint is flexed when the muscle contracts.

cle is fixed is more easily moved than that part on which it pulls by its other tendon ; the less movable attachment is the *origin*, the other the *insertion* of the muscle. Taking the biceps of the arm, we find that when the belly of the muscle contracts and pulls on its upper and lower tendons, the result is commonly that only the forearm is moved, the elbow joint being bent as shown in Fig. 29.

How may its thickening be recognized? What change besides thickening and shortening occurs in the belly of a contracted muscle? Give an example.
What is meant by the origin of a muscle? What by the insertion? Give an example.

The shoulder is so much more firm that it serves as a fixed point, and so that end of the biceps is the origin of the muscle, and the radial attachment its insertion. The distinction is, however, only relative : if the radius were held immovable the muscle would move the shoulder towards the radius, instead of the radius towards the shoulder ; as, for example, in going up a rope " hand over hand."

Varieties of Muscles.—Many muscles have the simple typical form of a belly tapering towards each end, as *A*, Fig. 30 ; others divide at one end, and are called two-headed, or *biceps* muscles, and there are even three-headed or *triceps* muscles. On the other hand, some muscles have no tendon at all at

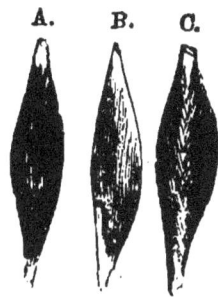

Fig. 30.—Diagrams illustrating, A, typical muscle with a central belly and two terminal tendons. B, a penniform muscle; C, a bipenniform muscle.

one end, the belly running right up to the bone to which it is fixed, and some have no tendon at either end. Sometimes a tendon runs along the side of a muscle, and the fibres of the latter are attached to it obliquely (*B*, Fig. 30) ; such a muscle is called *penniform* or feather-like, from a fancied resemblance to the vane of a feather ; or a tendon may run

Fig. 31.—A digastric muscle.

down the middle of the muscle (*C*), which is then called *bipenniform*. Sometimes a tendon is found in the middle of the belly as well as at each end (Fig. 31) ; such a

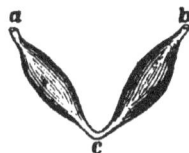

Is the origin of a muscle under all circumstances its most fixed end? Give an example.

What is the simple typical form of a muscle? What is a biceps muscle? What a triceps? Have all muscles tendons at each end? At either end? Describe a penniform muscle. A bipenniform.

muscle is called *two-bellied* or *digastric.* Running along the front of the abdomen, from the pelvis to the chest, on each side of the middle line, is a long muscle, *the straight muscle of the abdomen (rectus abdominis) ;* it is *polygastric,* consisting of four bellies separated by short tendons. Many muscles are not rounded, but form wide, flat masses, as those which lie beneath the skin on the sides of the abdomen.

How the muscles are controlled.—Most of the muscles of the body are paired in a double sense. In the first place, to nearly every one answers a corresponding muscle on the opposite side of the body,* its true mate ; in addition, most are paired with, or rather pitted against, an antagonist ; for example, to the biceps muscle (Fig. 29) which lies in front of the humerus and bends the elbow joint, corresponds the triceps muscle which lies behind the arm bone and extends the elbow ; when the biceps contracts the triceps relaxes, and *vice versa.* This orderly working is carried out by means of the brain and spinal cord, which, through the nerves, govern the muscles and regulate their activity. In convulsions these controlling organs are out of gear, and the muscles are excited to contract in all sorts of irregular and useless ways ; antagonists pulling against one another at the same moment the whole body is made rigid.

A digastric. Where do we find a polygastric muscle? How is the rectus abdominis muscle constituted? Where are flat wide muscles found ?

In what two ways are muscles paired? Give an example of antagonistic muscles. What happens to the triceps when the biceps contracts? How is the orderly working of the muscles guided and controlled? What parts are out of working order in a fit of convulsions? Why do the limbs often become stiff in convulsions?

---

* The single muscles cross the middle line and are made up of similar right and left halves; examples are orbicularis oris and the diaphragm.

The Gross Structure of a Muscle.—Each muscle is an organ composed of several tissues. Its essential constituent is a number of fibres consisting of *striped muscular tissue.* These are supported and protected by connective tissue; intertwined with blood and lymph vessels, which convey nourishment and carry off waste matters; and penetrated by nerves which govern their activity.

A loose sheath of connective tissue, *the perimysium,* envelopes the whole muscle in a sort of case; from it partitions run in and subdivide the belly of the muscle into bundles or *fasciculi* which run from tendon to tendon, or the whole length of the muscle when it has no tendons. The coarseness or fineness of meat depends on the size of these fasciculi, which may be readily seen in a piece of

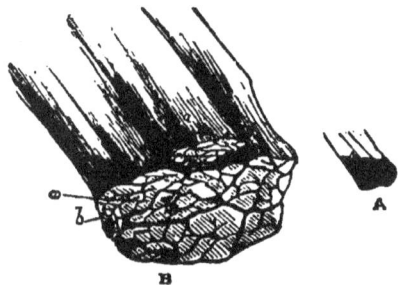

FIG. 32.—A small bit of muscle composed of four primary fasciculi. *A,* natural size; *B,* the same magnified, showing the secondary fasciculi of which the primary are composed.

boiled beef. In good carving, meat is cut across the fasciculi, or "across the grain," as it is then more easily broken up by the teeth; the polygonal areas seen on the surface of a slice of beef are cross sections of the fasciculi. The larger fasciculi are subdivided by fine partitions of connective tissue into smaller (Fig. 32), each consisting of a few *muscular fibres* enveloped in a close

Is a muscle an organ or a tissue? What is the chief tissue in it called? What things exist in it besides striped muscular tissue? What is the use of each?

What is the perimysium? How is a muscle divided into fasciculi? How far do the fasciculi extend? When is meat coarse in texture? Why is beef carved across the grain? Of what are the fasciculi composed?

network of minute blood-vessels.   Where a muscle tapers the muscle fibres in the fasciculi are less numerous and when a tendon is formed they disappear altogether, leaving only the connective tissue.

Histology of Muscle.—*The striped muscular tissue*, which gives the muscle its power of contracting, is found when examined by the microscope to be made up of extremely slender muscular fibres, each about one inch in length, but most of them less than $\frac{1}{500}$ of an inch across.

Each muscular fibre has externally a thin sheath or envelope, the *sarcolemma*, which envelops the contracting part of the fibre.   This latter is soft and almost semi-fluid, and under a microscope is seen to present a striped appearance, as if made up of alternating dimmer and brighter transverse bands (Fig. 33). After death the semi-solid contents of the fibre solidify and death-stiffening is produced ; at the same time the fibre often splits up into a number of very fine threads or *fibrillæ*, which were formerly regarded as true constituents of the living muscular fibre.

FIG. 33.—A small piece of muscular fibre highly magnified.   At *a* the fibre has been crushed and twisted so as to tear its contents, while the tougher sarcolemma, elsewhere so closely applied to the rest as to be invisible, remains untorn and conspicuous.

Plain muscular tissue.—The muscles hitherto spoken of

Of what is a tendon made ?

Of what is striped muscular tissue composed ? Describe the form and size of muscular fibres ?

What is the *sarcolemma?*   What is the consistency of the contractile part of a living muscular fibre?   What appearance does it present under the microscope? What is the cause of death stiffening ? What are fibrillæ ?

What do we mean by voluntary muscles ?

are all more or less under the control of the will ; we can make them contract or prevent this as we choose ; they are therefore often called the *voluntary* muscles.* There are in the body other muscles whose contractions

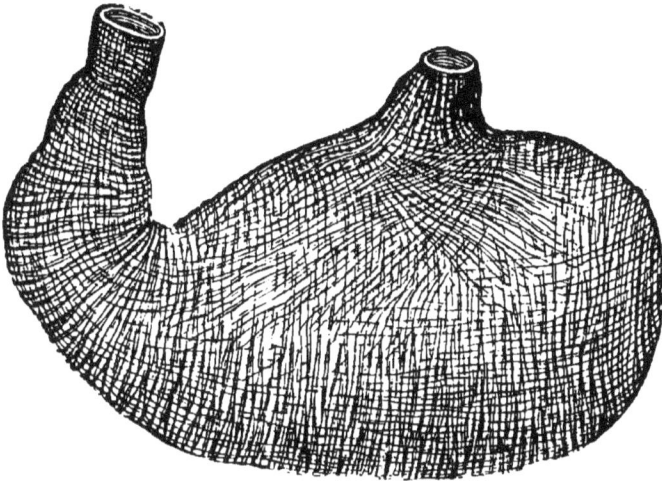

Fig. 34.—The muscular coat of the stomach.

we cannot control, and which are hence called involuntary muscles ; they are not attached to the skeleton directly, nor concerned in our ordinary movements, but lie in the

What by involuntary? Which kind is attached to the skeleton? Where do we find the involuntary muscles?

---

* No sharp line can be drawn between voluntary and involuntary muscles ; the muscles of respiration are to a certain extent under the control of the will ; any one can draw a long breath when he chooses. But in ordinary quiet breathing we are quite unconscious of their working, and even when we pay heed to it our control of them is limited : no one can hold his breath long enough to suffocate himself. Indeed, any one of the striped muscles may be thrown into activity, independently of or even against the will, as we see in the " fidgets " of nervousness, and the irrepressible trembling of extreme terror. Functionally, when we call any muscle voluntary, we mean that it may be controlled by the will, but not that it necessarily always is so. Structurally, the heart occupies an intermediate place : its striped fibres resemble much more those of voluntary than of involuntary muscles, but its beat is not at all subject to the will ; though, as the exception proving the rule, it may be noted that there is an apparently well-authenticated case of a person who could by an act of will stop his heart.

walls of various hollow organs of the body, as the stomach (Fig. 34), the intestines, and the arteries; by their contractions they move things contained in those cavities Like the voluntary muscles, the involuntary consist of contractile elements, with accessory connective tissue, blood-vessels, and nerves; but their fibres have a very different appearance under the microscope. They are not cross-striped, but are made up of elongated cells united by a small amount of cementing material. Each cell (Fig. 35), is flattish, and tapers off toward its ends; in its centre is a nucleus with one or two nucleoli. The cells have the power of shortening in the direction of their long axes.

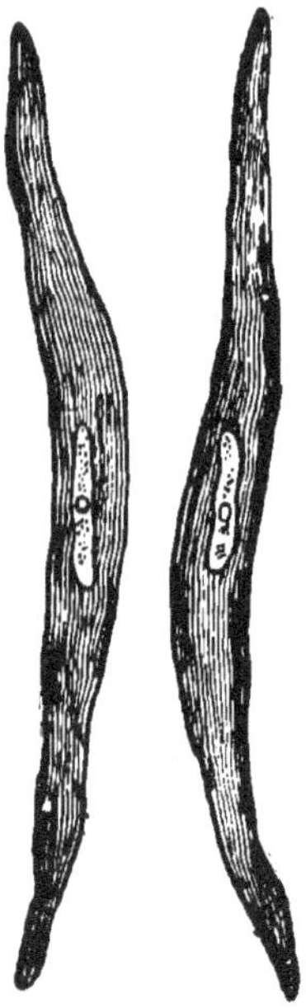

Heart muscle.—The muscular tissue of the heart is not under the control of the will; it, however, is cross-striped, and more like the voluntary than the ordinary involuntary muscle, though it differs in some respects from both.

FIG. 35.—Unstriped muscle-cells.

Speaking generally, we may say that the movements necessary for the nutrition of the body are not left for us to look after ourselves, but are carried on by muscles which work involuntarily; the blood is pumped round by the heart, and food churned up in the

What is their function? What are they composed of? What is seen when a cell from an involuntary muscle is examined with the microscope?

Is the heart muscle voluntary? In what respect does it resemble voluntary muscle?

What movements of the body does nature not leave to our own control? Give examples.

stomach and passed along the intestines, whether we think about it or not.

The chemical composition of muscle.—Muscle contains about 75 per cent. of water ; and a considerable quantity of salines. Living, resting muscle is alkaline to test paper ; hard-worked or dying muscle is acid. Its chief organic constituents are proteid or albuminous substances (p. 21), and of these the most abundant in a perfectly fresh muscle is *myosin*. Soon after death the myosin clots. Dilute acids dissolve myosin and turn it into syntonin, which used to be thought the chief proteid of muscle.

Beef tea.—When lean meat is heated its myosin is converted into a solid insoluble substance much like the white of a hard-boiled egg. Hence, when a muscle is boiled most of its proteid is coagulated and stays in the meat instead of passing out into the soup. Even if beef be soaked first in cold water this is still the case, as myosin is not soluble in water.* It follows that beef tea as ordinarily made contains little but the flavoring matters and salts of the beef, and some gelatin dissolved out from the connective tissue of the muscle. The flavoring matters

What proportion of water does muscle contain? What other inorganic compounds do we find in it? What is the reaction of living muscle? How is this changed by work or death? What are its main organic constituents? Name the most abundant of these? What change occurs in it after death? What is syntonin?

What happens to the myosin when muscle is heated? When we boil meat does its myosin become dissolved in the soup? Can we get the myosin out of beef by soaking it in cold water? What things are found in ordinary beef tea?

---

* To get over this difficulty, various methods of making beef tea have been suggested, in which the chopped meat is soaked an hour or two in strong brine or in very dilute muriatic acid. In these ways the myosin can be dissolved out of the beef; but the product has such an unpleasant taste that no one is likely to swallow it, and least of all a sick person,

make it deceptively taste as if it were a strong solution of the whole meat, whereas, it contains but a small proportion of the really nutritious parts, which are chiefly left behind in tasteless shrunken shreds, when the liquor is poured off. Some things dissolved out of the meat make beef tea a stimulant to the nervous system and the heart, but its nutritive value is small, and it cannot be relied upon to keep up a sick person's strength for any length of time.

Liebig's extract of meat is essentially but a concentrated beef tea; from its stimulating effect it is often useful to persons in feeble health, but other food should be given with it. It contains all the flavoring matters of the meat, and its proper use is for making gravies and flavoring soups; the erroneousness of the common belief that it is a highly nutritious food cannot be too strongly insisted upon, as sick persons may be starved on it if ignorantly used.

Various *meat extracts* are now prepared by subjecting beef to chemical processes in which it undergoes changes like those experienced in digestion. The myosin is thus made soluble in water and uncoagulable by heat, and a real concentrated meat extract is obtained. Before relying on any one of them for the feeding of an invalid, it would, however, be well to insist on having a statement of its

Why does beef tea taste as if all the "strength" of the meat were in it? Where do the chief nutritious parts remain when beef tea is strained off the meat? What is the action of beef tea on the system?

What is Liebig's extract of meat? Why is it sometimes useful to invalids? What should be given them in addition? What is its proper use? Why is it important to know that it is not a nutritious food?

How are some other meat extracts made? How is the myosin changed in preparing them? Are they all to be relied on indiscriminately? What should be done before trusting the nutrition of a feeble person to any one of them?

method of preparation, and then to consult a physician, or some one else who has the requisite knowledge, in order to ascertain if the method is such as might be expected to really attain the end desired.

CHAPTER VIII.

MOTION AND LOCOMOTION.

The special physiology of muscles.—The distinctive properties of muscle are everywhere the same ; it has the power of contracting ; but the uses of different muscles are very varied by reason of the different parts to which they are attached. Some are *muscles of respiration,* others *of swallowing ;* some bend joints and are called *flexors,* others straighten them and are called *extensors,* and so on. The determination of the exact use of any particular muscle is known as its *special physiology,* as distinguished from its *general physiology,* or properties as a muscle, without reference to its use as a muscle in a particular place. We may here consider the special physiology of the muscles concerned in standing and walking.

Levers in the body.—In nearly all cases the voluntary muscles carry out their special functions with the co-operation of the skeleton ; most of them are joined to bones at each end and when they contract move the bones,

In what respect are all muscles alike? Have all muscles the same uses? Give instances of the employment of muscles for different purposes. What is meant by the special physiology of a muscle? What by its general physiology?

With what do the voluntary muscles co-operate? To what are the ends of nearly all muscles attached? What happens when a muscle contracts?

and, secondarily, the soft parts attached to these. When muscles move bones the latter are almost invariably to be regarded as levers whose fulcra lie at the joint where the movement takes place. Examples of the three forms of levers recognized in mechanics are found in the human body.

Levers of the first order.—In this form (Fig. 36), the fulcrum or fixed supporting point, F, lies between the weight to be moved and the moving power. The distance

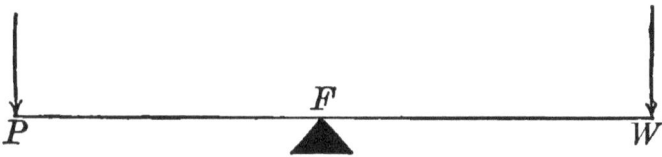

FIG. 36.—A lever of the first order. *F*, fulcrum ; *P*, power ; *W*, resistance or weight.

PF from the power to the fulcrum is called the power-arm of the lever, and the distance *WF* is the weight-arm. When power-arm and weight-arm are equal (as in an ordinary pair of scales), no mechanical advantage is gained ; to lift a pound at *W*, *P* must be pressed down with a force greater than a pound ; and the end *W* will go up just as far as the end *P* goes down. If *PF* be longer than *WF* then a small weight at *P* will balance a larger one at *W*, the gain being greater the greater the difference in the length of the arms, but the distance through which *W* is moved will be less than that through which *P* moves ; for example, if *PF* be twice as

How are the bones moved by muscles to be regarded? Where do the fulcra of these levers lie? How many kinds of levers are found in the body?

Describe a lever of the first order. Define power-arm and weight-arm. When is a mechanical advantage gained by such a lever?

long as WF then half a pound at P would balance against a pound at W, and just over half a pound laid on the end P would lift a pound on the end W, but W would only go up half as far as P went down. On the other hand, if the weight-arm were longer than the power-arm there would be a loss in force, but a gain in the distance through which the weight was moved.

Examples of levers of the first order are not numerous in the human body. One is found in nodding movements of the head, the fulcrum being where the occipital bone articulates with the atlas (Fig. 20). When the chin is raised the power is applied to the skull behind the fulcrum by muscles passing from the spinal column

Fɪɢ. 37.—A lever of the second order. F, fulcrum; P, power; W, weight. The arrows indicate the direction in which the forces act.

to the back of the head ; the resistance to be overcome is the excess in weight of the part of the head in front of the fulcrum over that behind it, and is not great, as the head is nearly balanced on the top of the spine. To let the chin drop does not necessitate any muscular effort.

Levers of the second order.—In this form of lever (Fig. 37), the weight or resistance acts between the fulcrum and the power. The power-arm PF is accordingly

What is lost when power is gained?
Are there many levers of the first order in the body? Give an example of one, describing the action.
Describe a lever of the second order.

always longer than the weight-arm, WF, and so a comparatively weak force can overcome a considerable resistance. There is, however, a loss in rapidity and extent of movement, since it is obvious that when P is raised a certain distance W will be raised less. As an example of this kind of lever we may take the act of standing on the toes. Here the foot is the lever, and the fulcrum is where its fore part rests on the ground ; the weight is that of the body, and acts downwards through the ankle joint at Ta, Fig. 19 ; the power is the great muscle of the calf of the leg pulling by its tendon, which is fixed to the end of the heel bone, Ca.

Levers of the third order.—In these (Fig. 38), the power is applied between the fulcrum and the weight ; hence the power-arm PF, is always shorter than the

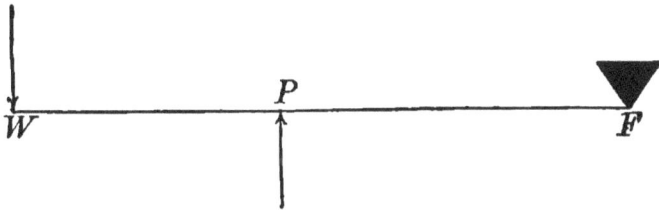

Fig. 38.—A lever of the third order. F, fulcrum ; P, power ; W, weight.

weight-arm, WF. The moving force acts at a mechanical disadvantage, but swiftness and range of movement are gained ; this is the form of lever most commonly used in the body. For example, when the forearm is bent up towards the arm the fulcrum is the elbow joint (Fig. 29) ;

What is the mechanical gain in such levers?   What is the loss? Give an example of employment of a lever of the second order in the body, pointing out fulcrum, point of action of the weight, and point of application of the power.

Describe a lever of the third order.   What is lost and what gained by it?   Is it often used in the body?   Give an example

the power is applied at the insertion of the biceps **muscle**
into the radius; the weight is that of the forearm **and**
hand and whatever may be held in the latter, and acts **at**
the centre of gravity of the whole, somewhere on the far
side of the point of application of the power. Usually
(as in this case), the power-arm is very short, so as to gain
speed and extent of movement, the muscles being strong
enough to work at a considerable mechanical disadvantage.
The limbs are thus also made much more shapely than
would be the case were the power applied near or beyond
the weight.

Pulleys in the body.—Fixed pulleys are used in the
body; they give rise to no loss or gain of power, but
serve to change the direction in which certain muscles
pull. One of the muscles of the eye-ball, for example,
has its origin at the back of the eye-socket, from
there it passes to the front and ends, before it reaches the
eye-ball, in a long tendon. This tendon passes on to the
margin of the frontal bone, which arches over the front
of the eye-socket, and there passes through a ring
and turns back to the eye-ball. The direction in which
the muscle moves the eye is thus quite different from
what it would be if the tendon went directly to the eyeball.

Standing—We only slowly learn to stand in the year
or two after birth, and though we finally come to do it
without conscious attention, standing always requires the
co-operation of many muscles, guided and controlled by

Why is the power-arm in the body usually short?
What kind of pulley is used in the body? Is any mechanical ad-
vantage gained from it? What is it used for? Give an example.
Is standing a simple process?

the nervous system. The influence of the latter is shown by the fall which follows a severe blow on the head, which has fractured no bone and injured no muscle; "the concussion of the brain" stuns the man, and until it has passed off he cannot stand.

When we stand erect, with the arms close by the sides and the feet together, the centre of gravity of the whole adult body lies at the articulation between the sacrum and the last lumbar vertebra, and a perpendicular drawn from it will reach the ground between the feet. In any position in which this perpendicular falls within the space bounded by a line drawn close around both feet, we can stand. When the feet are together the area enclosed by this line is small, and a slight sway of the trunk would throw a perpendicular dropped from the centre of gravity of the body outside it; the more one foot is in front of the other the greater the sway back or forward which will be compatible with safety, and the greater the lateral distance between the feet the greater the lateral sway which is possible without falling. Consequently, when a man wants to stand very firmly he advances one foot obliquely, so as to increase his base of support both from before back, and from side to side.

In consequence of the flexibility of its joints a dead body cannot be balanced on its feet as a statue can. When we stand, the ankle, knee, and hip-joints, if not braced by the muscles, would give way, and the head also

Illustrate the influence of the nervous system in connection with standing.

Where is the centre of gravity of the body when we stand erect? Where does a perpendicular from it reach the ground? Why do we separate the feet when we want to stand firmly?

Why cannot a dead body be balanced on its feet? What prevents our knee, and hip-joints from bending when we stand?

fall forward on the chest. But (Fig. 39) muscles, 1, in front of the ankle-joint, and others, I, behind it, both contracting at the same time, keep the joint from yielding; similarly muscles (2) in front of the knee and hip-joints are opposed by others (II) behind them, and when we stand both contract to a certain extent and keep those joints rigid; and the muscles (III), which run from the pelvis to the back of the head similarly pull against others, 3 and 4, which run from the pelvis to the lower end of the breastbone, and from the upper end of the breastbone to the anterior part of the skull, and their balanced contraction keeps the head erect. Since the degree to which each muscle concerned contracts when we stand must be accurately adjusted to the contraction of its antagonist on the opposite side of the joint, we may easily comprehend why it takes us some time to learn to stand, and why a stunned man, whose muscles have lost guidance from the nervous system, falls.

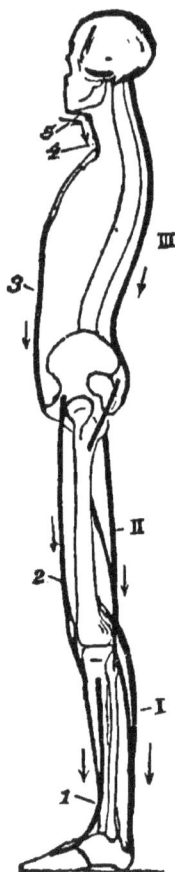

FIG. 39.—Diagram illustrating the muscles (drawn in thick black lines) which pass before and behind the joints, and by their balanced activity keep the joints rigid and the body erect.

Locomotion includes all movements of the body in space, dependent on its own unaided muscular efforts, such as walking, running, leaping, and swimming.

Explain how the different joints concerned are "braced" in standing. Why does it take a child some time to learn to stand?
What is meant in physiology by locomotion?

Walking.—In walking, the body never entirely quits the ground, the heel of the advanced foot reaching this before the toe of the rear foot has been raised from it. In each step the advanced leg supports the body, and the foot behind at the beginning of the step propels it.

A little attention will enable any one to analyze the act of walking for himself. Stand with the heels together and take a step, commencing with the left foot. The whole body is at first inclined forwards, the movement taking place mainly at the ankle joints. This throws the centre of gravity in front of the base formed by the feet, and a fall would result were not the left foot simultaneously raised by bending the knee a little, and swung forwards, the toes just clear of the ground and the sole nearly parallel to it. When the step is completed the left knee is straightened and the foot placed on the ground, the heel touching first ; the base is thus extended in the direction of the stride and the fall prevented. Meanwhile the right leg is kept straight but inclined forwards, carrying the trunk during the step while the left foot is off the ground ; at the same time the right foot is raised, commencing with the heel ; when the step of the left leg is completed only the great toe of the right is in contact with the support. With this toe a push is given which sends the body swinging forward, supported on the left leg, which now in turn is kept rigid except at the ankle joint ; the right knee is immediately afterwards bent and that leg swings forwards, its foot just clear of the ground, as the left did before. The body meanwhile is supported

Is the body ever off the ground in walking? Describe the act of walking.

on the left leg alone. When the right leg completes its step its knee is straightened and the foot thus brought, heel first, on the ground ; while it is swinging forwards the left foot is gradually raised, and at the end of the step its great toe alone is on the ground ; with this a push is given as before with that of the right foot, and the left leg then swings forward to make the next step. Walking may, in fact, be briefly described as the act of continually falling forwards and preventing the completion of the fall by thrusting out a leg to meet the ground in front.

During each step the body sways a little from side to side, as it is alternately borne by the right and left legs. It also sways up and down a little ; a man standing with his heels together is taller than when one foot is advanced, just as a pair of compasses held erect on its points is higher when its legs are together than when they straddled apart ; in that period of each step when the advancing trunk is balanced vertically over one leg, the walker's trunk is more elevated than when the front foot also is on the ground. Women, accordingly, often find that a dress which clears the ground when they are standing sweeps the pavement when they walk.

The length of each step is primarily dependent on the length of the legs, though it can be largely controlled by special muscular effort, as we see in a regiment of soldiers, all of whom have been taught to take the same stride, no matter how their legs vary in length. In natural easy walking, little muscular effort is employed to carry the rear leg forward after it has given its push; it swings on

At what part of a step is a man tallest ? Give illustrations.
What primarily determines the length of a person's step ? Can this length be controlled ? Illustrate.

like a pendulum once its foot is raised from the ground. As short pendulums swing faster than long ones the natural step of short-legged people is quicker than that of long-legged.

Running differs from walking in several respects. There is a moment when both feet are off the ground ; the toes alone come in contact with it at each step ; and the knee joint is not straight at the end of the step. In running, when the rear foot is to leave the ground the knee is suddenly straightened, and the ankle-joint extended so as to push the toes forcibly on the support and powerfully impel the whole body forwards and upwards. The knee is then considerably flexed and the foot raised some way from the ground, and this occurs before the toes of the front foot reach the support. The raised leg in each step is forcibly drawn forward by its muscles and not allowed to swing passively as in quiet walking. This increases the rate at which the steps follow one another, and the stride is increased by the sort of one-legged jump that occurs through the jerk given by the straightening knee of the rear leg, just before it leaves the ground.

Hygiene of the Muscles.—The healthy working of the muscles is dependent on a healthy state of the body in general ; this is indispensable that they may be sufficiently supplied with proper nourishment, and have their wastes promptly carried away. Hence good food and pure air are necessary for a vigorous muscular system. Muscles also

Why do short-legged persons tend to take a quicker step than others?

How does running differ from walking? Describe the act of running. How is the number of steps taken in a given time increased in running? How is the stride increased?

How does the state of general health influence the muscular system? Why does an athlete need good food and air?

should not be exposed to any considerable continued pressure, since this interferes with the flow of blood and lymph through them which is essential for their nutrition.

Exercise is necessary for the best development of the muscles. A muscle long left unused diminishes in bulk and degenerates in quality, as is well seen when a muscle is paralyzed and remains permanently inactive because of disease of its nerve ; although at first the muscle itself may be perfectly healthy, it alters in a few weeks, and when the nerve is repaired the muscle may in turn be incapable of activity. The same fact is illustrated by the feeble and wasted state of the muscles of a limb which has been kept motionless in splints for a long time : when the splints are removed it is only after careful and persistent exercise that the long idle muscles regain their former size and power. The great muscles of the "brawny arm" of the blacksmith illustrate the converse fact—the growth of muscles when exercised.

Exercise, to be useful, must be judicious ; taken to the point of extreme fatigue, day after day, it does harm. When a muscle is worked its substance is used up ; at the same time and afterwards more blood flows to it, and if the exercise is not too violent and the intervals of rest are long enough, the repair and growth will keep pace with or exceed the wasting : but excessive work and too short rest will lead to diminution and enfeeblement of the muscle just as certainly as too little exercise.

Few persons can profitably attempt to work *hard* daily

Why should muscles not be exposed to continuous pressure?
What happens when a muscle is not used? Illustrate by examples.
Why are the muscles of a blacksmith's arm large?
When does exercise do harm? Why? Can most persons work hard with both brain and muscle at the same time?

with both brain and muscle, but all should regularly use both; choosing which to *work* with, and which to simply exercise. The best earthly life, that of the healthy mind in the healthy body, can only so be attained. For persons of average physique, engaged in study or business pursuits of a sedentary nature, the minimum of daily exercise should be an amount equivalent to a five-mile walk.

Time for Exercise.—Since extra muscular work means extra muscular waste, and should be accompanied by an abundant supply of food materials to the muscles, violent exercise should not be taken after a long fast. Neither should it be taken immediately after a meal; a great deal of blood is then needed in the digestive organs to provide materials for digesting the food, and this blood cannot be sent off to the muscles without the risk of an attack of indigestion. Strong and hearty young people may take a long walk before breakfast, but others had better wait until after eating something before engaging in any kind of hard work.

Varieties of Exercise.—In walking and running the muscles chiefly employed are those of the lower limbs and trunk; these exercises leave the muscles of the chest and arms imperfectly worked. Rowing is better, since in it nearly all the muscles are used. No one exercise employs in proper proportion all the muscles, and gymnasia in which different feats of agility are practiced so as to call different muscles into action have a deserved popu-

How can the highest development of man, regarded merely as a thinking and moving machine, be attained?

Why should we not exercise when fasting? Why not soon after eating?

What muscles are chiefly used in walking and running? Which are imperfectly exercised? Why are gymnasia useful?

larity. It should be borne in mind, however, that the legs especially need strength; while in the arms delicacy of movement is more important to most persons than great strength; and the fact that gymnastics are usually practised indoors is also a great drawback to their value. Out-of-door exercise in good weather is better than any other, and every one can at least take a walk. The daily "constitutional" is very apt to become wearisome, especially to young persons, and exercise loses half its value if unattended with feelings of mental relaxation and pleasure. Active games, for this reason, have a great value for young and healthy persons; lawn-tennis, base-ball, and cricket are all attended with pleasurable excitement, and are excellent also as exercising many muscles.

Effects of Alcohol on Muscle.—It has been found by experiment that in most cases the drinking of quantities of wine or beer which do not intoxicate will, nevertheless, diminish the muscular effort which a healthy person can exert: he can no longer raise the weight which he could before. Such experiments have been carefully made on men under military regimen; they proved that alcohol does not increase the power of sustained muscular work, though it may for a brief time excite to unhealthy activity.

Another frequent result of "moderate" drinking is *tremor,* or shakiness of the hands. The hand is unsteady even when the arm is folded, and is seen to tremble if it be held out with the arm extended. This unsteadiness interferes with the performance of any action calling for muscular precision or manual dexterity. Another drink is apt

What is chiefly needed in the muscles of arms and legs respectively? Point out conditions under which muscular exercise loses much of its value. Why are athletic games especially useful?

to be resorted to "to steady the hand." Entire abstinence from alcohol, the essential cause of the unsteadiness, is the only real remedy. The extra glass may for a short time steady the muscles, but, repeated, leads to the acquirement of serious diseases.

Total abstinence from alcohol and tobacco is required from all competitors while in training for athletic games and races. A man who, after election as a member of his college crew, should be found secretly drinking beer or smoking, would be hissed out of college. So, in later years, in the race for success in life, the abstainer has the advantage.

Deterioration of Tissue due to Alcohol.— A serious structural change in the body produced by alcohol is *fatty degeneration.* The oily matter of the body exists in two forms: first, as adipose or fatty tissue collected under the skin, and in less amount elsewhere, as on the surface of the heart and around the kidneys; second, as minute fat-droplets in the interior of various cells and fibres. Some forms of alcoholic drinks tend to increase the adipose tissue, and may lead to undue accumulation of it about the heart, which is a muscle, impeding the action of that organ. A more important and frequent result is an increase of fat-droplets in the cells and the muscular fibres, the oily matter replacing the natural working substance.

What is the effect of alcoholic drinks on muscular power? On muscular precision? How does alcohol lead to deterioration?

CHAPTER IX.

WHY WE EAT AND BREATHE.

How is it that the body can do muscular work?—In the muscles we possess a set of organs capable of moving the body from place to place, of changing the relative positions of its parts, and of lifting external objects: as long as we are alive, more or fewer of our muscles are every moment doing some mechanical work. This fact suggests the question, where does this power of working come from? In a few words, the answer is, it comes from the burning of parts of the body itself: in the burning, work-power or *energy* is set free and some of this is used by the muscles.

The conservation of energy.—The different natural forces known to us are not nearly so numerous as the kinds of matter: we all, however, know several of them, as light, heat, electricity, and mechanical work. One of the greatest discoveries of the nineteenth century is that these different natural forces, or *forms of energy,* can be turned one into another, directly or indirectly: kinds of energy are transmutable, while, so far as we know at pres-

What are the functions of the muscles? Are all of our muscles ever at rest at the same time?

What question does the constant activity of our muscles suggest? How may this question be briefly answered?

Name some forms of energy. Can they be turned from one form to another?

ent, kinds of matter are not. We cannot, as the alchemists hoped, turn iron or mercury into gold, but we can turn light into heat, and heat into electrical force, or into mechanical work. When such transformations are made it is always found that a definite amount of one kind of energy disappears to give rise to a certain definite amount of another. In other words, it has been discovered that energy cannot be created : if we take a given quantity of heat we can turn some of it into mechanical work ; if we then turn all this mechanical work back into heat we get again exactly the quantity of heat which disappeared when the mechanical work appeared : and so with all other transformations of energy from one kind to another, and back again. This fact that *energy or work-power can be turned from one kind into another, and often back again, but never created from nothing or finally destroyed*, is known as *the law of the conservation of energy.* .

Illustrations of the conservation of energy.—In a steam-engine, heat, which is the best known kind of energy, is produced in the furnace ; when the engine is at work all of this energy does not leave it as heat ; some is turned into mechanical work, and the more work the engine does the greater is the difference between the heat generated in the furnace and that leaving the machine. If, however, we used the work to rub two rough surfaces together we could get the heat back, and if (which of course is impossible in practice) we could avoid all friction between

Can matter be transmuted ? What is always found when energy is transformed ?

Can man create energy ? Illustrate the fact that energy can be changed in kind but not created. What is meant by the law of the conservation of energy ?

Give an illustration of the conservation of energy.

the moving parts of the machine, and have all parts of the engine at the end of the experiment exactly at the same temperature as at its beginning, the quantity of heat thus obtained would be exactly equal to the difference between that amount of heat originally generated in the furnace of the engine, and the quantity which had been carried off from it to the air since its fire was lighted. Having turned some of the heat into mechanical work we could thus turn the work back into heat again, and find it yield exactly the amount which seemed lost.

Or we might use the engine to drive an electro-magnetic machine and so turn part of the heat liberated in its furnace, first into mechanical work, and this afterwards into electricity; and if we chose to use the latter with the proper apparatus, as now used for electric lighting, we could turn more or less of it into light; and so have a great part of the energy which first became conspicuous as heat in the engine furnace, now manifested in the form of light at some distant point. In fact, starting with a given quantity of one kind of energy, we may by proper contrivances turn all or some of it into one or more other forms; but if we collected all the final forms and retransformed them into the first, we should have exactly the amount of it which had disappeared when the other kinds appeared.

Why we need food.—Energy, as we have seen, cannot be created from nothing; since the body constantly expends energy, it must have a steady supply. This supply comes

Give an example of the transformation of heat into electrical force. Of electrical force into light. Given a supply of one kind of energy what can we do with it? What would we find if we collected all the final manifestations of energy and turned them back into the original form?

Why must the body have a steady supply of energy? Where does the supply come from?

from the energy liberated when parts of the body are burned, or, as the chemists say, *oxidized*, just as that used by a locomotive comes from the burning or *oxidation* of coal or wood in its furnace.   In consequence of this constant oxidation, which destroys the tissues of the body as coal is destroyed in a furnace, new materials must constantly be supplied to make up for those used for oxidation.   These new materials are provided in our *food*.   One chief reason of our needing to eat, is that we may replace the parts of the body which have been burned in order to set free the energy which we spend in our muscular movements.

Why the body is warm.—As a working steam-engine is warm so are our bodies, because all the energy which is set free when substances are burned in them, is not turned into mechanical work, but some of it appears as heat. This keeping warm is a very important matter, for experiment shows that no tissue of the human body works well when cooled down even a few degrees below 98.5° F., which is its natural healthy temperature.   Careful experiments prove that when a muscle does work it becomes hotter, and we all know that exercise makes us warm. This shows that the oxidation or burning which takes place in a working muscle does not all become turned into mechanical work, but a good share of it appears as heat. What is true of muscle is true of all other organs of the body: when they work, no matter what their kind of work, their substance is oxidized, and some of the energy set free

What is the chemical term for burning? What does food supply? Point out a chief reason for our need of eating?

Why are our bodies warm ?   Why is it important that they should be warm ?   How is the temperature of a muscle affected when it works ?   Do other organs resemble muscles in this respect?

by the oxidation appears as heat, assisting to keep the body warm, and at its best working temperature.

A second reason why we need food.—Since the body only works well at a temperature which is higher than that of the air around it (except on a very hot day), and in health always keeps at this temperature, it must lose heat nearly all the time. At night each of us is, in health, just as warm as in the morning; and in the morning as when we went to bed; though we have lost heat to the air during the day, and to the bedclothes at night. In order to keep our bodies at the temperature most suitable to their activity, they must, therefore, generate heat all the time, to compensate for the giving of it from them to the outer world. In this necessity of generating heat we find a second reason for the need of food: we require daily to take into ourselves things which can be burned (or oxidized) in the body, and which in so doing will give off heat.

The influence of starvation upon muscular work and animal heat.—When a man is deprived of food the supply of things which can be oxidized in his body is cut off. The tissues and organs are used up and not renewed; his temperature falls, his muscles become weaker and weaker, and at last he dies. The body does not live, and work, and keep warm, by means of a peculiar vital force or energy which inhabits it, but by utilizing the energy set free in it by the oxidation of foods, or of things made in it from foods. If the food supply be cut off, the body first uses up

What must we conclude from the fact that our bodies keep at nearly the same temperature all the time? How do we know that they generate heat? Give a reason for taking food, in addition to its use as a source of energy to be spent by the muscles.

What happens when a man is starving? How does the body live, and keep warm, and work? What first happens when the food supply is stopped?

any reserve of nutritious matter which may have been stored up in it when the starvation commenced, and as this is expended it becomes weaker and weaker until death supervenes.*

How long a man, totally deprived of food, can keep alive, will depend, partly, on how much reserve material, capable of oxidation, he has stored up in him when the starvation period commences; but largely, also, on the extent to which he can spare himself muscular work and loss of heat. The breathing movements and beat of the heart must go on, but if the individual lies quiet in bed he need do little or no other muscular work; and if he is well covered up with blankets, the loss of heat from the body is slight and calls for but little oxidation of the tissues to compensate for it.† Also, a fat person will survive starvation longer than a lean one; during the process his fat is slowly burnt; but so long as it lasts he can supply his muscles with something which can be oxidized to yield working power, and he also, by its burning, can maintain his temperature. Fat is, in fact, a sort of reserve fuel, laid up in the body, and a man, in the strict sense of the word, can hardly be said to begin to *starve* until his fat has nearly all been used up.‡

Upon what does the length of life of a man getting no food depend? What expenditures of energy must go on all the time? How does lying in bed diminish the expenditure of energy? Why will a fat man deprived of food live longer than a lean one?
When does a fasting man really begin to *starve?*

---

* When warm-blooded animals are starved their temperature slowly falls; and when it comes down to about 77° F. (25° C.) death occurs: the various tissues at that temperature can no longer work so as to maintain life.

† Hence Dr. Tanner, and "fasting girls" keep in bed, warmly covered up, most of the time: the losses of the body in mechanical work and heat are thus reduced to a minimum, and consequently the oxidation of the food reserves stored in the body at the beginning of the fast.

‡ Some warm-blooded animals, as bears, *hibernate;* that is, sleep all through

Oxidations in the body.—In the preceding paragraphs oxidation and burning have been used as equivalent phrases : this is in accordance with the teachings of chemistry. To the chemist a substance is *burned* when it is combined with oxygen, whether this combination take place slowly or rapidly. If the combination occur rapidly the burning or *oxidizing* mass becomes very hot and also gives off light: such a rapid and vigorous oxidation is called a *combustion ;* no combustions take place in our bodies.

It has, however, been proved that whether the combination of oxygen with an *oxidizable*, or *burnable*, substance takes place rapidly or slowly, at the end of the process exactly the same amount of energy will have been set free in each case. When the oxidation occurs in a few seconds the oxidizing mass becomes very hot : when it occurs slowly, in a few days or weeks, the mass will never be very hot, because the heat set free in the process is carried off nearly as fast as it appears.

Illustrations of oxidations at a low temperature.—If a piece of magnesium wire be ignited in the air it will become white-hot, flame, and leave at the end of a few

What does a chemist mean when he says a substance is burned ?
What is a combustion ?   Do combustions occur in our bodies ?
Does the quantity of energy liberated by the complete oxidation of any substance vary with the rate of oxidation ?
Why is a slowly oxidizing mass of matter not very hot ?
Give an instance of the oxidation of the same substance at high and low temperatures.

---

the winter and take no food.   They feed well in the warm weather, and are quite fat at the close of autumn, when they seek some shelter d place to winter in.   This shelter and their warm, furry coats make the loss of heat very little; the animal, except for its breathing and the beat of its heart, hardly ever moves during the winter, and even those necessary movements are reduced to the least possible, the breathing and heart-beat being much slower than during the summer.   With return of warm weather the creature wakes up again, but is then lean, having burnt up its fat during its winter sleep.

seconds only a certain amount of incombustible *rust* or *magnesia*, which consists of the metal combined with oxygen ; under these circumstances it has been burnt or *oxidized* quickly at a high temperature. The heat and light evolved in the process represent the energy which is set free by the metal and oxygen when they combine. We can, however, oxidize the metal in a different way, attended with no evolution of light and no very perceptible rise of temperature If, for instance, we leave it in wet air, it will become gradually turned into magnesia without having ever been hot to the touch or luminous to the eye. The process then, however, takes days or weeks ; but in this slow oxidation just as much energy is liberated as in the former case, although now all takes the form of heat ; and instead of being liberated in a short time is spread over a much longer one, as the gradual chemical combination takes place. The slowly oxidizing magnesium is, in consequence, at no moment noticeably hot, since it loses its heat to surrounding objects almost as fast as it generates it. The oxidations occurring in our bodies are of this slow kind. An ounce of arrowroot oxidized in a fire, and in the human body, would liberate exactly as much energy in one case as the other, but the oxidation would take place in a few minutes and at a high temperature in the former, and slowly, at a lower temperature, in the latter.

Oxidation in the presence of moisture.—Wet wood or wet coal we know will not burn, or can only be made to do so with difficulty. Other kinds of burning or oxidation are, however, well known, which take place in the pres-

How does the rate of oxidation differ in the two cases ? How does the oxidation of arrowroot burned in a fire differ from its oxidation in the living body?
Can oxidations occur in the presence of moisture ?

ence of moisture. The rusting of iron, for example, is an oxidation or burning of the metal, and takes place faster in damp air than in dry; during the slow rusting in moisture just as much heat is set free as if the same compound of iron and oxygen were prepared in a more rapid way. Such experiments throw great light on the oxidations which take place in our own bodies. All of them are slow oxidations, which never at any one moment give off a great amount of heat, and all occur in the damp tissues.

Summary. (1) Energy can be turned from one form into another ; as from heat into mechanical work by a steam-engine. (2) Our bodies are constantly losing energy, partly in muscular work, and partly as heat lost to surrounding objects. (3) Energy cannot be created ; all that can be done is to turn one kind of it into another kind: heat can be turned into mechanical work (as in a locomotive) ; or mechanical work into heat (as by friction) ; or heat into electricity (as in a thermo-electric machine); and so forth. (4) Since our bodies spend energy all our life long they must be supplied with it from outside : they can turn into other forms the energy which they receive, but they cannot make it from nothing. (5) The chief forms of energy which our bodies expend are muscular (*i. e.* mechanical) work, and heat. (6) In ordinary machines, as a locomotive, the

Give an instance. Does the rate of oxidation or the presence of moisture affect the amount of heat liberated ?

Of what kind are the oxidations which occur in our bodies ?

Give a summary of the contents of this chapter with reference to the following points: (1) The transformation of energy; (2) The loss of energy from the body ; (3) The fact that man cannot create energy but can transmute it ; (4) The fact that our bodies must be constantly supplied with energy from outside ; (5) The chief forms of energy spent by the body ; (6) The source of the energy spent by a working steam-engine.

source of the work done and the heat given off is the oxidation of coal in a furnace. (7) Chemistry teaches us that just the same amount of energy or work power is given off when an ounce of any given substance is oxidized, whether the oxidation occurs rapidly or slowly. (8) Chemistry also teaches us that many oxidations, or burnings, occur in the presence of water, and that in them just the same amount of energy is set free as if the oxidation occurred in dry air. (9) In our bodies substances are burnt slowly at a low temperature and in the presence of moisture: in this burning energy is set free which the body uses for performing its necessary work. (10) In the burning which the tissues undergo while they work they are used up and destroyed. (11) To compensate for the destruction of tissue which accompanies and provides the power for every action of the body, material must be taken into it from outside, which will restore or repair the oxidized tissues. (12) Such substances taken into the body from outside are called *foods,* and the constant oxidation of the body which is necessitated by the performance of the functions essential to life, requires a supply of food from the outer world, if life is to be maintained. (13) The body of a healthy person has in it at any moment a certain reserve of oxidizable matters, which we may call stored-up food. The most important of these reserve foods is fat. A fat man can accordingly bear starvation longer than a lean one under similar circumstances.

(7) The amount of energy given off when a substance is oxidized at high or low temperatures; (8) The teachings of chemistry with reference to oxidations in the presence of moisture ; (9) The conditions under which substances are oxidized in our bodies ; (10) The changes which oxidized tissues undergo ; (11) Why we need to take material from the outer world into our bodies ; (12) What is meant by *food;* and why we need foods ; (13) What are reserve foods. Illustrate.

The oxygen food of the body.—Hitherto we have only considered the energy-supply of the body from one side ; we have regarded it as dependent on the constant supply of things which can be oxidized. But this is only half the question : if substances are to be oxidized there must be a provision of oxygen to oxidize them.

In order that a steam-engine may work and keep warm it is not merely necessary that it have plenty of coal, but it must also have a draught of air through its furnace. Chemistry teaches us that the burning in this case consists in the combination of a gas called oxygen, taken from the air, with other things in the coals: when this combination takes place a great deal of heat is given off. The same thing is true of our bodies ; in order that food matters may be burnt in them and enable us to work and keep warm, they must be supplied with oxygen; this they get from the air by breathing. We all know that if his supply of air be cut off a man will die in a few minutes. His food is no use to him unless he gets oxygen from the air to combine with it ; while he usually has stored up in his body an excess of food matters which will keep him alive for some time if he gets a supply of oxygen, he has not stored up in him any reserve, or, if any, but a very small one, of oxygen, and so he dies very rapidly if his breathing be prevented. In ordinary language we do not call oxygen a food, but restrict that name to the solids and liquids which we swallow : but inasmuch as it is a material which we must take from the external universe into our bodies in or-

Why do we need oxygen ? What does a working steam-engine need in addition to coal? What happens in the furnace of an engine? Why do we need to breathe ? What happens if a man's air supply be stopped ? Why does a man die sooner of want of air than of want of food ? Why is oxygen entitled to be called a food ?

der to keep us alive, oxygen is really a food as much as any of the other substances which we take into our bodies from outside, in order to keep them alive and at work. *Suffocation,* as death from deficient air supply is named, is really death from oxygen-starvation.

What is suffocation?

Appendix to Chapter IX.

The liberation of energy by oxidation, or *burning*, at a low temperature and in the presence of moisture, is such a fundamental fact in physiology, and its essential agreement with ordinary combustion so difficult to grasp by most pupils, who naturally associate burning with a high temperature and luminosity, that it is worth while to illustrate these facts by a few simple experiments.

1. Buy a coil of magnesium wire, which can be obtained at small cost. Rub it clean with fine emery paper : cut it in two, apply a lighted match to one half and show how it is rapidly consumed with the evolution of light and heat, leaving behind only a white powder, magnesia, which is oxidized magnesium. Put the other half away in a bottle with a few drops of water. After a day or two its surface will be covered by a layer of magnesia: if this be scraped off another will succeed it ; and so on. This experiment shows that oxidation may occur rapidly at a high temperature in a short time, or slowly at a low temperature in a long time, but the ultimate product, in each case, is the same.

2. In relation to a subsequent paragraph (p. 132) the magnesia obtained by burning the wire in the air, may be kept, and attempts made to ignite it: this will serve to show the uselessness of oxidized substances to the body, as sources of energy : they cannot be any more oxidized, and the best thing to do is to get rid of them.

3. Get a bundle of iron wire ; rub it bright with emery or sand paper. Place some in a warm dry bottle by the stove or fireplace. Put the rest in a bottle containing a little water. Next day the first specimen will be found bright, and the second covered with rust. This shows that oxidation may sometimes occur better in the presence of water than in its absence ; and serves as a text for pointing out how oxidations occur in the moist tissues of the body.

CHAPTER X.

NUTRITION.

The Wastes of the Body.—A man takes into his body daily several pounds of foods of various kinds, as meats, bread, vegetables, and water, yet he grows no heavier; it is, therefore, clear that his body must in every twenty-four hours return, on the average, to the outside world about as great a weight of matter as it receives from it. Even in childhood, while growth is taking place and the body becoming heavier, the gain is never nearly equal to the weight of the foods swallowed. The materials given off daily from the body to the external universe, and compensating more or less accurately for the receipts from the outside world, are its *wastes*, and are chiefly things which cannot be burned. Much of the food taken in can be, and is, oxidized to enable us to move and keep warm. When burned it is of no further use to us, and would only clog up the various organs, as the ashes and smoke of an engine would soon put its fire out if they were allowed to accumulate in the furnace. The chief wastes of the body are *carbon dioxide gas*, *water*, and a kind of ammonia called *urea*.

Receptive and Excretory Organs.—Those organs of the body whose function it is to gather new material from outside for its use are known as *receptive organs*. There

What facts make it clear that a man must daily give off several pounds weight of matter from his body? Does a child's increase in weight equal the weight of the food it has eaten? What is meant by the " wastes " of the body? How do most foods differ from wastes in regard to oxidation? Why must wastes be removed from the body? Name the chief wastes of the body.

What is meant by the receptive organs?

are two chief sets of these—one to receive oxidizable things, and the other to receive oxygen. The first set is represented by the alimentary canal, consisting of *mouth, gullet,* * *stomach,* and *intestines.* It takes in food and drink. The second set consists of the lungs, with the air passages leading to them ; their business, as receptive organs, is to absorb oxygen.

The organs whose duty it is to get rid of waste materials formed in the body are called *excretory organs.* The three most important excretory organs are the *lungs,* the *kidneys,* and the *skin;* the lungs pass carbon dioxide gas out to the air, and also water ; the kidneys get rid of urea and water ; and the skin, of water and a little urea.

The Intermediate Steps between Reception and Excretion.—Between the taking of oxidizable substances into our mouths and oxygen into our lungs, on the one hand, and the return of oxidized matters from our bodies to the surrounding world on the other, a great many intermediate steps take place. The alimentary canal (see Fig. 1) is a tube which runs through the body but nowhere opens into it ; so long as food lies in this tube it therefore does not really form a part of the body, and is of no use to it : it resembles coals in the tender of a locomotive, waiting to be transferred to the furnace. In our bodies the furnace is everywhere; wherever there is a living tissue things are

What are the functions of the two chief sets? Name those concerned in receiving oxidizable things. Those whose business it is to absorb oxygen.

What is meant by the excretory organs? Name the most important. What does each get rid of?

Why is food in the stomach not really a part of the body? To what may we liken it? Where is the furnace of the body? Why must food be carried all over the body?

---

* The technical name for the gullet is *œsophagus.*

burned to enable it to work; and the food or fuel must be brought . therefore, to every corner of our frames.

Digestion.—A great part of our food is solid, and could not of itself get outside of the alimentary canal. To render it available it must be dissolved so that it can soak through the walls of the stomach and intestines. For this purpose we find a set of *digestive organs* which make solvent juices and pour them on the food which we swallow, and so get it into a liquid state in which it can be absorbed.

Circulation.—The solution containing our digested food if it simply soaked through the walls of the alimentary canal, would be of no use to distant parts, as the brain, or the muscles of the limbs. We find, therefore, in the body a set of tubes containing blood, and called blood-vessels: the blood is driven through these by a pump, *the heart.* Much of the dissolved food passes into the blood-vessels of the alimentary canal, and from them is carried by other blood-vessels to every organ, no matter how remote. As the blood flows unceasingly, round and round in its vessels, from part to part, the organs concerned in moving and conveying it are called *circulatory organs*, and the blood-flow itself is known as *the circulation.*

Absorbents.—Some of the dissolved food is taken up into another set of tubes in the walls of the alimentary canal; these tubes carry it afterwards into the blood-vessels. They are called *the absorbents.*

Why must many foods be dissolved? What is accomplished by the digestive organs?

What are the blood-vessels? What enters those of the alimentary canal? How are organs distant from the alimentary canal nourished? Why are the organs which keep up the blood-flow called circulatory organs? What is meant by the circulation?

What are the absorbents? Where do they convey the foods which they take up in the walls of the alimentary canal?

Respiration.—The blood in its course flows through **the** lungs. It is necessary not merely that food, but oxygen also should be carried to every part of the body. As the blood traverses the lungs it picks up oxygen from the air in them; this air is then renewed by taking a fresh breath, and so on. The organs concerned in renewing the air in the lungs are the *respiratory organs,* and the act of renewal is *respiration.*

Assimilation.—As each organ works it oxidizes; some of its substance is broken down by combination with oxygen brought to it by the blood, and is thus converted into burnt waste matter. The blood, as we have seen, brings, however, not merely oxygen, but also food matters in solution. These ooze through the walls of the blood-vessels, and are taken up by the living tissues and built into new tissues like themselves, to replace the part which has been used up and destroyed. This building and repair of tissues and organs from the dissolved food obtained from the blood is known as *assimilation,*—in plain English, "a making alike." Each living tissue takes from the blood foods which are not like itself, and builds them up into a form of matter like its own. The converse process, which accompanies all vital action, the breaking down into wastes of a living tissue when it works, is called *dissimilation,* or "a making unlike."

The Relation of the Circulatory Organs and the Absorbents to Excretion.—It is as essential to the body that its wastes be carried off from the organs, as that the used-up

What must be carried to all parts in addition to food? Where does the blood get oxygen? What is meant by the respiratory organs? By respiration?

What happens when an organ works? How are oxidized tissues replaced? What is meant by assimilation? By dissimilation?

What is needful to each organ in addition to a supply of fresh material?

material be replaced by new. Not merely must matter for assimilation be provided, but the various waste products must be removed. Here again the blood-vessels and absorbents come into play. Absorbents are found not only in the walls of the alimentary canal, but all over the body. The wastes of each working tissue are passed out into them, and by them carried into the blood-vessels; these in turn carry the wastes to the lungs, kidneys, and skin, which get rid of them. The blood is thus as important in relation to removing the waste matters of an organ as in regard to supplying it with food and oxygen.

Nutrition.—From what has been said above it is clear that the nourishment of the body is a very complicated process. It implies—(1) the reception of food from outside; (2) the digestion of food; (3) the absorption of digested food; (4) the conveyance of absorbed food to all parts by the blood; (5) the taking up of wastes from the different organs; (6) the conveyance of these wastes by the blood to excretory organs which pass them out of the body; (7) the absorption of oxygen in the lungs, and its conveyance by the blood to every organ; (8) assimilation or the building up of new tissue from materials brought by the blood; and (9) disassimilation, or the breaking down of working tissues by combination with oxygen.

In subsequent chapters we shall have to consider in more detail, Digestion, Circulation, Absorption, Respiration, and Excretion. The sum total of the actions of all the organs concerned in the nourishment of the body is known as the function of *nutrition.*

Where do we find absorbents in addition to those of the alimentary canal? What is their function? What part does the blood play in the removal of wastes?

Enumerate the processes concerned in the nourishing of the body. What is meant by the function of nutrition?

CHAPTER XI.

FOODS.

Foods as Tissue Formers.—In the last chapter we have considered foods merely as sources of energy, but they are also required to build up the substance of the body. From birth to manhood we increase in bulk and weight, and that not merely by accumulating water and such substances, but by forming more bone, more muscle, more brain, and so on, from the things which we eat. Even after full growth, when the body ceases to gain weight, the same constructive processes go on; the living tissues are steadily oxidized and broken down as they work, and as constantly reconstructed.

Foods are therefore needed, not only to supply the body with work-power by their oxidation, but to supply material from which new living tissue can be constructed.

What Foods must Contain.—Most foods serve for both purposes, energy supply and tissue formation; they are built up by the living cells into new tissue before they are oxidized to set energy free. Our food must, therefore, contain such substances as the body can utilize for tissue formation.* The living tissues when analyzed are found

What use have foods besides supplying energy to the body? Illustrate from the growth of a child. Why are foods needed for construction after growth has ceased?

What purposes do most foods serve? Are they usually oxidized before making tissue? What sort of substances must our food contain?

---

* Whether any food is ever oxidized in the body before being built up into a tissue, as coal is burnt in an engine without ever forming part of the engine, must still be regarded as an open question in physiology. The old doctrine that some foods, as starch and sugar, were useful only to set free heat, and others, as albumen and flesh, alone built tissue, must be given up. It seems certain

to consist mainly of carbon, hydrogen, nitrogen, and oxygen, and we might at first suppose that these chemical elements in their uncombined form would serve to nourish us. Experience, however, teaches that this is not the case. Four fifths of the air is nitrogen, but we cannot feed on it; hydrogen gas is of no use as a food; and a lump of charcoal (carbon) might fill the stomach, but would not keep a man from starving. Oxygen can be utilized when taken by the lungs from the air; but all other elements to be of use as food must be taken, not in their separate state, but in the form of complex compounds, in which they are chemically combined with other things; as, for example, in starch, and sugar, and fat, and oil, and albuminous substances.

The Special Importance of Albuminous or Proteid Foods.
—All the active tissues of the body are found to yield on chemical analysis large quantities of albumens. (See p. 21.) As the tissues work this proteid is broken down, and its nitrogen carried off in the form of a peculiar ammoniacal substance, *urea;* to repair the wasted living tissue new proteids must be laid down in it. So far as we know at present the human body (like that of most animals) is unable to make proteids out of other things; given one vari-

What elements do the tissues yield on analysis ? Can we feed on these elements in their uncombined state ? Name one which is absorbed in a free state and used. Whence is it derived ? What organs receive it? Name substances containing the necessary elements in combination and used as food.

What do we find in all the active tissues? What becomes of the nitrogen of working tissues? Explain why proteids are an essential article of diet.

---

that under some conditions sugar and starch may be used in building tissue, though they cannot do it alone; but whether they are under any circumstances ever burnt before making part of a tissue is not certain. On the other hand, there is some reason to suspect that albuminous substances may, when eaten in excess, be oxidized in the body without ever forming part of a living cell.

ety of them it can turn it into other varieties, but it cannot make proteids from things which are not proteids. Hence these albuminous or proteid substances are an essential article of diet.

The Limited Constructive Power of the Animal Body.— From what has been said above it is clear that our bodies are, on the whole, destructive rather than constructive in relation to the outer world. They require for their nutrition very complex chemical compounds (starch, sugar, fat, proteids), build these up into living tissues, and then oxidize the tissues and return the carbon, hydrogen, and nitrogen, which were received from outside in the form of complicated chemical molecules, to the outer world in the form of much simpler chemical compounds, namely, carbon dioxide, water, and urea. None of these latter substances is capable of nourishing an animal ; it cannot from them alone build up its tissues or set free energy.

How Plants Supply Food for Animals, and Animals Food for Plants —Since animals are essentially proteid consumers, and destroyers also of other complex substances, as starch and sugar, the question naturally suggests itself, How is it, if animals are constantly consuming these things, that the supply of them is kept up? For example, the supply of proteids ; they cannot be made artificially by any process known to us. The answer is, that animals live on the things which plants make, and plants live on

Do our bodies on the whole build up or break down chemical compounds? What class of compounds do they require for their nutrition? What do they do with these compounds? What simple compounds does the body return to the outer world? Can these compounds feed any animal?

What facts suggest the question, How is the supply of proteids and other complex foods kept up? How is the question answered?

the carbon dioxide and water and ammonia (urea) which animals excrete.

As regards our own bodies the question might, indeed, be apparently answered by saying that we get our proteids from the flesh of the other animals which we eat. But, then, we have to account for the possession of proteids by those animals; since they cannot make them from urea and carbon dioxide and water any more than we can. The animals whose flesh is used by us as food get their proteids from plants, which are the great proteid formers of the world; the most carnivorous animal really depends for its most essential foods upon the vegetable kingdom; the fox that devours a hare, in the long run lives on the proteids of the herbs that the hare had previously eaten.*

Non-Oxidizable Foods.—Besides our oxidizable foods a large number of necessary food materials are not oxidizable, or at least are not oxidized in the body. Typical instances are afforded by water and common salt. The use of these is in great part physical: the water, for instance, dissolves materials in the alimentary canal, and carries the solutions through its walls into the blood and lymph vessels, so that they can be conveyed from place to place; and it permits interchanges by enabling the things it has dissolved to soak through the walls of the vessels. The salines also influence the solubility and chemical interchanges of other things present with them. *Serum albumen,* one of the proteids

Where does the proteid that a man eats in a piece of beef come from? Explain.

What foods are necessary in addition to oxidizable? Give examples. What are their physical uses?

---

* Some animals are known which contain chlorophyl, the green coloring matter of plant leaves; and it has recently been proved that these animals, like plants, can, when exposed to the action of light, live on the waste products of other animals.

which is carried in the blood all over the body to supply albuminous material to the tissues, is, for example, insoluble
in pure water, but dissolves readily if a small quantity of
common salt be present. Besides such uses, the non-oxidizable foods have probably others, as what may be called
machinery formers. In the lime salts, which give their
hardness to the bones and teeth, we have an example of such
an employment of them; and to a less extent the same may
be true of other tissues. The body is a self-building and
self-repairing machine, and the material for this building
and repair, as well as the fuel or oxidizable foods which
yield the energy the machine expends, must be supplied in
the food. While experience shows us that even for
machinery construction oxidizable matters are largely
needed, it is nevertheless a gain to replace such substances
by non-oxidizable material when possible; just as, if practicable, it would be advantageous to construct an engine
out of a substance which would not rust, although other
conditions determine the selection of iron for building the
greater part of it.

Definition of Foods.—*Foods are* (1) *substances which are
taken into the alimentary canal, which can be absorbed
from it, and after absorption serve to supply material for
the growth of the body, or for the replacement of matter
which has been removed from it; or* (2) *they are substances
which can be oxidized in the body to yield energy for its use;
or* (3) *substances, which by dissolving nutritive or waste*

Illustrate the use of common salt in helping to keep important
substances in solution. Illustrate the employment of non oxidizable
foods in constructing the body. What must foods supply to the
body besides fuel? Why? Are oxidizable foods used in machinery
construction? Give an example showing the gain of using non-
oxidizable matters when possible.
   Give a definition of foods.

*matters facilitate the transfer of material from the recep-
tive organs to the working, and from the working to the
excretory.* Foods to replace matters which have been oxidized must be themselves oxidizable; they are *force gener-
ators*, but may be and generally are also tissue formers: they
are nearly always complex organic substances derived from
other animals or from plants. Foods to replace matters
not oxidized in the body, as water and salt, are *force regulators*, and are for the most part tolerably simple inorganic
compounds. Among the force regulators we must, however, include certain foods, which, although oxidized in the
body and serving as sources of energy, yet produce effects
totally disproportionate to the amount of energy which they
thus set free. Their influence as stimulants in exciting certain tissues to activity, or as agents checking the activity of
parts, is more marked than their direct action as force gener-
ators. As examples, we may take condiments: mustard and
pepper are not of much use as sources of energy, although
they no doubt yield some when oxidized; we take them for
their stimulating effect on the mouth and other parts of the
alimentary canal, by which they promote a greater flow
of the digestive secretions or an increased appetite for food.
Thein, again, the active principle of tea and coffee, is taken
for its stimulating effect on the nervous system rather than
for the amount of energy which is yielded by its own oxi·
dation.

To the above definition of a food should be added the
condition that, *neither the substance itself nor any of the*

What foods must be oxidizable? What are they called? Do they
also make tissue? Are they complex or simple? What is their source?
What is meant by force regulators? Examples? Are they chemically complex or simple? What oxidizable foods are included
among the force regulators? How is their influence chiefly exhibited? Give examples.

products of its chemical transformation in the body shall be injurious to the structure or action of any organ; otherwise it would be a poison, not a food.

Alimentary Principles.—The substances which in ordinary language we call foods are in nearly all cases mixtures of several *foodstuffs* with substances which are not foods at all. Bread, for example, contains water, salts, gluten (a proteid), some fats, much starch, and a little sugar; all these are true foodstuffs, but mixed with them is a quantity of *cellulose* (the chief chemical constituent of the walls which surround vegetable cells), and this is not a food, since it is incapable of absorption from the alimentary canal. Chemical examination of all the common articles of diet shows that the actual number of important *foodstuffs* is but small; they are repeated in various proportions in the different foods we eat, mixed with small quantities of different flavoring substances, and so give us a pleasing variety in our meals; but the essential substances are much the same in the fare of the artisan and in the "delicacies of the season." The chief foodstuffs, which are found repeated in many different foods, are known as "*alimentary principles*," and the nutritive value of any article of diet depends on the proportion of these foodstuffs which is present, far more than on the various agreeable flavoring matters which cause certain things to be especially sought after, and to have a high market value. Alimentary principles may be conveniently classified as albumens, albuminoids, hydrocarbons, carbohydrates, and inorganic bodies.

What is a poison?
What are ordinary foods? Give an example. Why is cellulose not a food? What does chemical examination of ordinary foods show? How do we get variety in our foods? What are "alimentary principles"? On what does the nutritive value of a food depend? Into what groups are alimentary principles classified?

Albuminous Alimentary Principles.—Of the nitrogen-containing foodstuffs the most important are albumens: they are an essential part of all diets, and obtained both from animals and plants. The most common and abundant are myosin and syntonin, which exist in the lean of all meats; egg albumen; casein, found in milk and cheese; gluten and vegetable casein from various plants.

Albuminoid Alimentary Principles.—These also contain nitrogen, but cannot entirely replace the proteids as foods; though a man can manage with less albumen when he has some albuminoids in addition. The most important is *gelatine*, which is yielded by the connective tissue and bones of animals when cooked. On the whole, albuminoids are not foods of high value, and the calf's-foot jelly and such compounds often given to invalids have not nearly the nutritive value they are commonly supposed to possess.

Fats and Oils, or Hydrocarbons.—The most important of these are stearin, palmatin, margarin, and olein, which exist in various proportions in animal fats and vegetable oils; the most fluid containing most olein : butter contains a special fat known as butyrin. All fats are compounds of glycerine with fatty acids, and, speaking generally, any such substance which is fusible at the temperature of the body will be useful as a food. The stearin of beef and mutton fats is not by itself fusible at the body temperature, but is mixed in those foods with so much olein as to be melted

What nitrogen-containing substances form an essential article of diet? Whence are they obtained? Name the most important albuminous foods.

What foods besides albumens contain nitrogen? To what extent can they take the place of albumens? How is gelatin obtained? Should we try to build. up an invalid's strength on calf's-foot jelly alone? Give the reasons for your answer.

To what group of foods do fats and oils belong? Name the more important. What is their chemical composition? Why are some fatty bodies not nutritious? Give an instance.

in the alimentary canal. Beeswax is a fat which does not melt in the intestines and so is unabsorbed, although from its composition it would be useful as a food could it be digested.

Starchy and Sugary Foods are often named **Carbohydrates.**—They are mainly of vegetable origin. The most important are *starch*, found in nearly all vegetable foods; *dextrin; gums; grape sugar* (found in most fruits); and *cane sugar. Sugar of milk* and *glycogen* are alimentary principles of this group derived from animals. All carbohydrates, like the fats, consist of carbon, hydrogen, and oxygen; but the percentage of oxygen in them is much higher than in fats. When oxidized they have therefore less power of combining with additional oxygen than fats, and so are not capable of yielding as much energy to the body.

Inorganic Foods.—The most important of these are water; common salt; and the chlorides, phosphates, and sulphates, of potassium, magnesium, and calcium. A sufficient quantity of most of these substances, or of the material for their formation, exists in all ordinary articles of diet, so that we do not swallow most of them in a separate form. Water and table salt form exceptions to the rule that inorganic bodies are eaten imperceptibly along with other things, since the body loses more of each daily than is usually supplied in that way. It has been maintained that salt as a separate article of diet is an unnecessary luxury, and there seems to be some evidence that certain savage tribes live without more than they get in the meat and vegetables

What is the chief source of carbohydrates? Name the most important. Name carbohydrate foods derived from animals. How do they resemble fats in composition, and how do they differ from them? Can an ounce of starch yield as much energy to the body as an ounce of fat? Give a reason for your answer.

Name the more important inorganic foods. Why do we not require to eat most of them separately? What inorganic foods are taken in a separate form? Why?

which they eat. Such tribes are, however, said to suffer especially from intestinal parasites; and there is no doubt that to many animals as well as most men the absence of salt from their diet is a terrible deprivation. Buffaloes and other creatures are well known to travel miles to reach " saltlicks;" of two sets of oxen, one allowed free access to salt, and the other given none save what existed in its ordinary food, it was found after a few weeks that those given salt were in much better condition. In man the desire for salt is so great that in regions where it is scarce it is used as money. In some parts of Africa a small quantity of salt will buy a slave, and to say that a man commonly uses salt at his meals is equivalent to stating that he is a luxurious millionaire. In British India, where the poorer natives regard so few things as necessaries of life that it is hard to levy any excise tax, a large part of the revenue is derived from a salt tax, salt being something which even the poorest will buy. As regards Europe, it has been found that youths in the Austrian empire who have fled to the mountains and there led a wild life to avoid the hated military conscription, will, after a time, though able abundantly to supply themselves with other food by hunting, come down to the villages to purchase salt, at the risk of liberty and even of life.

The Nutritive Value of Different Foods.—All *meats*, whether derived from beast, bird, or fish, are highly valuable foods. They contain abundant albumen, more or less fat, and, when cooked, their connective tissue is in great part turned into gelatin. *Pork* is the least easily digested form of fresh meat, and contains a larger percentage of fat than most. This fat, which, by its oxidation liberates much heat, makes it a good food in cold weather for persons with a good digestion. Pigs are especially liable to a parasite, called *trichina*, which lives in their muscles, and may be transferred thence to man, sometimes causing death. Hence pork should always be thoroughly cooked. Salted meats of all kinds are less digestible and less nutritious than fresh. *Milk* contains an albuminous substance (casein), also fats (butter), and a sugar, known as *sugar of milk*, in addition to useful mineral alimentary principles. It will support life longer than any other single food. *Cheese* consists essentially of the casein of milk: it is a very nutritive albuminous food. *Eggs* contain albumens and fats, and have a high nutritive value: they are more easily digested when cooked soft than hard. *Wheat* contains more than a tenth of its weight of proteids, more than half its weight of starch, some sugar, and a little fat. The albumen of wheat flour is mainly *gluten*, which when moistened with water forms a tenacious mass, and this it is to which wheaten bread owes its superiority. When the dough is made, yeast is added to it and causes fermentation by which, among other things, carbon dioxide gas is produced.* This gas, imprisoned in the tenacious dough and expanded by heat during baking, forms cavities in it, and causes the dough to "rise" and make "light bread," which is not only more pleasant to eat but more easily digested than heavy. Some

Why should pork be well cooked? How do salted meats differ in value as articles of diet from fresh? What foodstuffs exist in milk? What one food will support life longest? What is the chief alimentary principle in cheese? What is the nutritive value of cheese? What makes the value of eggs as foods? Are they more easily digested soft or hard boiled? What foodstuffs exist in wheat? Why is wheaten bread lighter than that made from other grains?

* The small amount of alcohol generated in the fermentation or rising of bread readily passes off in the heat of the oven. There is no alcohol in wellbaked bread.

grains contain a larger percentage of starch, but none have so much gluten as wheat ; when bread is made from them the carbon dioxide gas escapes so readily from the less tenacious dough that it does not expand the mass properly. *Corn* contains less albumen, more starch, and more fat than wheat. *Rice* is poor in albumen but very rich in starch. *Peas* and *beans* are rich in albumen and contain about half their weight of starch. *Potatoes* are not very nourishing, because they contain a great deal of water and only about one pound of albumen and fifteen pounds of starch in every hundred pounds of potatoes. Other fresh vegetables, as carrots, turnips, and cabbages, are valuable mainly for the salts they contain; their weight is chiefly due to water, and they contain but little starch, albumens, or fats. *Fruits*, like most fresh vegetables, are mainly valuable for their saline constituents, the other foodstuffs in them being only present in small proportion. Some kind of fresh vegetable is, however, a necessary article of diet, as shown by the scurvy which used to prevail among sailors before fresh vegetables or lime-juice were supplied to them.

Alcohol.—We have learned already that alcohol-containing drinks are apt to injure the joints and muscles ; and in subsequent chapters their hurtful influence on nervous, and digesting, and circulating, and breathing organs will be pointed out. For the present we confine ourselves to the question, Has alcohol a just claim to be called a food ? Does it build tissue, or strengthen the muscles, or help to maintain our animal heat ? Is it useful to health ?

How does corn differ from wheat in composition? What alimentary principles are scarce in rice? Which one is abundant? What do potatoes contain? What is the main useful constituent of most fresh vegetables? Of fruits? How may scurvy be prevented?

Why are all alcoholic drinks dangerous? What questions must be answered before deciding if alcohol is a true food ?

Is Alcohol a Tissue-Forming Food?—To this the answer is certainly *no ;* so far at least as useful tissue is concerned. Alcohol cannot build up albuminous material, since it contains no nitrogen ; and such material constitutes the essential part of muscular, glandular, and nervous tissues. There is even some evidence that alcohol leads to excessive waste of such tissues : several competent observers have found that its use increases the amount of nitrogen waste excreted from the body. It often leads to excessive and harmful overgrowth of connective tissue and fat, but it does not lead to development of muscle or brain or gland.

Is Alcohol a Strengthening Food?—To this the answer is also *no.* Alcohol in small doses excites brain and muscle, and may for a short time goad them to overwork or to work when they should be resting. But as it nourishes neither of them, the final result is bad. The brain and muscle are left in an injured state. As regards the brain, the consequence is often insanity (Chap. XXI.). As regards the muscles, very careful experiments have been made on soldiers who were given definite tasks to accomplish. The result was that on the days on which they were supplied with spirits, they could neither use their muscles as powerfully, nor for as long a time, as on the days when they got no alcoholic drink.

Does Alcohol keep up the Heat of the Body?—To this question, also, the answer is *no,* though this may seem strange in view of the fact that a drink is often taken "to warm one up." The apparent inconsistency is easily

What is said of alcohol as a tissue-forming food?
Is alcohol a strengthening food? How may it lead to overwork? Results? What were the results of experiments made on soldiers as to the action of alcohol on the muscles?
Does alcohol maintain the heat of the body? Why does a drink sometimes make a person feel warmer?

explained. Our feeling of being warm depends on the nerves of the skin (p. 376). We have no nerves which tell us whether heart or muscles or brain are warmer or cooler. These inside parts are always hotter than the skin, and if blood which has been made hot in them flows in large quantity to the skin, we feel warmer because the skin is heated. As alcoholic drinks make more blood flow through the skin, they often make a man feel warmer. But their actual effect upon the temperature of the whole body is to lower it. The more blood that flows through the skin, the more heat is given off from the body to the air, and the more blood, so cooled, is sent back to the internal organs. The consequence is that alcohol, in proportion to the amount taken, cools the body as a whole, though it may for a short time heat the skin. That a large dose of alcohol leads to excessive loss of heat from the body has been proved by many observations on drunken men, and by experiments on the lower animals.

The study of alcohol as an article of diet leads therefore to the result that it cannot fairly be regarded as a food.

Tea, Coffee, and Cocoa are excitants rather than nutritive foods. The amount of nourishment in a cup of either is but little. Some persons experience wakefulness or a feeling of fulness in the head after taking coffee, and such should of course avoid it. Sportsmen out for a long day's shooting find cold tea superior to spirits ; ¦military commanders find a ration of coffee far better than one of whiskey for fatigued troops, and all arctic explorers have come to a similar conclusion.

How is it that alcohol sometimes makes a person feel warmer? How does it cool the body?

To what class of foods do tea and coffee belong? What results do they produce? Why are they better than alcohol for similar purposes? Give illustrations of the influence of tea and coffee in removing the sensation of fatigue.

Cooking.—When meat is cooked most of its connective tissue is turned into gelatin, and the whole mass becomes softer and more readily broken up by the teeth. In boiling meat it is a good plan to put it first into boiling water which coagulates its surface layer of albumen, and this then keeps in flavoring and other matters which would otherwise pass out into the water. After the first few minutes the cooking should be continued at a lower temperature; meat boiled too fast is hard, tough, and stringy. In roasting or baking meat, the same plan is advisable. Put it close to the fire or in a hot oven for a short time, and then complete the cooking more slowly at a lower temperature.

The cooking of vegetable foods is of considerable importance. Starch is the chief nutrient matter in most of them, and raw starch is much less easily digested than cooked. When starch is roasted it is turned into a substance known as *soluble starch,* which is easily dissolved by the digestive liquids, so there is a scientific foundation for the common belief that the crust of a loaf is more digestible than the crumb, and toast than fresh bread.

The Oxidizable Matters required daily by the Body.— The necessary quantity of daily food depends upon that of the material used by the body and passed out from it in each twenty-four hours; this varies both in kind and amount with the work done and the organs most used. In children a certain excess is required to furnish material for growth.

What happens when meat is cooked?

Why does a good cook first put meat that she is to boil into very hot water? Why should the boiling be completed at a lower temperature? How should meat be baked?

Why is it important to cook most vegetable foods? Why is toast more easily digested than fresh bread?

What determines the necessary amount of daily food? How does it vary? Why do children require more in proportion to their size than adults?

It is impossible to state accurately beforehand just what amount of food any individual will require, but a general idea may be arrived at by taking the average daily losses, by excretion, of a man, as determined by many experiments made on different persons. Such experiments show that a man of average size and doing ordinary work needs rather more than $9\frac{1}{2}$ ounces (274 grams) of carbon to replace his loss of that element; and about $\frac{7}{10}$ of an ounce of nitrogen (20 grams). Some hydrogen is also required, as the body daily loses more water than we take in our food; and this extra amount implies a loss of hydrogen, combined with oxygen in the body to form water.

The Advantages of a Mixed Diet.—Since albuminous foods contain carbon, nitrogen, and hydrogen, life may be maintained on them if the necessary salts, water, and oxygen be also supplied; but such a diet would not be economical. Albumens contain in 100 parts about 52 of carbon and 15 of nitrogen, so a man fed on them alone would get about $3\frac{1}{2}$ parts of carbon for every 1 of nitrogen. His daily losses are not in this ratio, but about 13.7 parts of carbon to 1 of nitrogen; and to get enough carbon from albumens far more than the necessary amount of nitrogen must be taken. Of dry albumens 1 pound $2\frac{1}{2}$ ounces (527 grams) would yield the necessary carbon, but would contain $2\frac{3}{4}$ ounces (79 grams) of nitrogen, or four times more than is necessary to cover the daily losses of that element from the body. Fed on a purely albuminous diet a man would, therefore, have to digest a vast quantity to get

What is the average daily loss of carbon from the body? Of nitrogen? Does a man need hydrogen also in his food? Why?

On what group of foodstuffs can life be maintained without any others? Why is feeding entirely on albuminous substances not desirable?   —

enough carbon, and in eating and absorbing it, and in getting rid of the excess nitrogen (which is useless to him), a great deal of useless labor must be thrust upon the digestive and excretory organs. Were a man to live on bread alone he would force much unnecessary work on his organs. Bread contains little nitrogen in proportion to its carbon, and to get enough nitrogen far more carbon than could be utilized would have to be eaten, digested, and excreted daily.

The human race has discovered this fact: men use, where they have a choice, richly proteid substances to supply the nitrogen needed, but derive the carbon mainly from non-nitrogenous foods of the fatty or starch and sugary kinds, and so avoid excess of either nitrogen or carbon. For instance, lean beef contains about $\frac{1}{4}$ of its weight of dry albumen, which albumen contains 15 per cent of nitrogen. Consequently 1 pound 3 ounces of lean meat would supply the nitrogen needed to compensate for a day's losses. But the albumen contains 52 per cent of carbon, so the amount of it in the above weight of fatless meat would be 1070 grains (69 grams) or nearly $2\frac{1}{2}$ ounces, leaving 3150 grains (205 grams) or rather more than seven ounces, to be got either from fats or sugary and starchy substances. The necessary amount would be contained in 3940 grains (256 grams) or about 9 ounces of ordinary fats, or in 7080 grains (460 grams), a little over a pound, of starch; hence either of these with the above quantity of

Explain why bread by itself would afford a bad diet.

Why do men use a mixed diet? Explain why lean meat alone would not be a good food. How could the deficient carbon of lean beef be supplied?

Give illustrations of the fact that most foods contain more than one foodstuff.

lean meat would form a far better diet both for the purse and the system than meat alone.

As already pointed out, nearly all common foods contain several *foodstuffs*. Good butcher's meat, for example, contains nearly half its dry weight of fat: and bread in addition to proteids contains starch, fats, and sugar. In neither of them, however, are the foodstuffs mixed in the physiologically best proportions, and the custom of consuming several of them at each meal, or different ones at different meals during the day, is not only agreeable to the palate but in a high degree advantageous to the body. The strict vegetarians who do not eat even such substances as eggs, cheese, and milk, but confine themselves to a purely vegetable diet, which is always poor in albumens, take daily far more carbon than they require, and are to be congratulated on their excellent digestions which are able to stand the strain. Those so-called vegetarians who use eggs, cheese, etc., can of course get on very well, since such substances are extremely rich in albumens, and supply all the nitrogen needed, without the necessity of swallowing the vast bulk of food which must be eaten in order to get it directly from plants.

Why do we commonly use several foods at one meal? What element do strict vegetarians take in excess? How do nominal vegetarians get their nitrogen?

CHAPTER XII.

THE DIGESTIVE ORGANS.

General Arrangement of the Alimentary Canal.—The alimentary canal is a tube which runs through the body from the lips to the posterior end of the trunk. It is lined by a soft reddish *mucous membrane* (easily seen inside the mouth), which is but a redder and moister sort of skin. Outside the mucous membrane are connective tissue and muscular layers, which strengthen the digestive tube and push the swallowed food along it. The mucous membrane is constructed to absorb dissolved nutritive substances; it soaks them up and passes them into blood or lymph vessels. Imbedded in this mucous membrane, or lying outside it, are hollow organs called *glands;* these glands make liquids which alter chemically many substances which we eat, and turn them from things which cannot be absorbed by the mucous membrane into things which can. The whole series of changes which any food material undergoes, between its reception by the mouth and its absorption by the alimentary mucous membrane, is spoken of as its *digestion*.

Various foodstuffs undergo different kinds of changes

What is the alimentary canal? By what is it lined? What are found outside the mucous membrane of the digestive tube? What are their uses? With reference to what object is the alimentary mucous membrane constructed? What does it do with the nutriment it absorbs? What is the function of the glands of the alimentary canal? What is meant by the *digestion* of a foodstuff?

preliminary to absorption, and so we speak of different kinds of digestions; as that of starch, of fats, of albuminous bodies, and so forth.

Glands are hollow organs which make or *secrete* peculiar fluids and pour them out on some free surface of the body. They are very widely distributed; we find, for example, digestive glands (of several kinds) opening into the digestive tube, perspiratory glands opening in the skin, tear glands or *lachrymal glands* pouring out their secretion on the eyeball. Different glands have their cavities lined by different kinds of cells, and produce different secretions. In general arrangement all glands are built on one or other of two primary structural plans, known as the *tubular* and the *racemose.*

The Kinds of Glands.—All portions of the body making and pouring forth secretions are not technically called glands. In the peritoneum, which lines the inside of the abdominal cavity (p. 11), we find simply a thin membrane (*A*, Fig. 40), having on its side nearer the cavity which it surrounds a layer of cells, *a*, and on its deeper side a network of very fine blood-vessels, *c*, supported by connective tissue, *d*. Such simple, smooth, secreting surfaces are not common; in most cases an extended area is required to form the necessary amount of secretion, and if this were attained simply by spreading out flat membranes, these, from their number and extent, would be hard to pack conveniently in the body. Accordingly, in most cases, a large area is obtained by folding the secreting surface in various

Why do we speak of different kinds of digestion? Illustrate. What is a gland? Give examples of glands. How do glands differ? What are the two chief types of glands named?

Name and describe a secreting surface which is not technically called a gland. What is gained by folding a secreting surface?

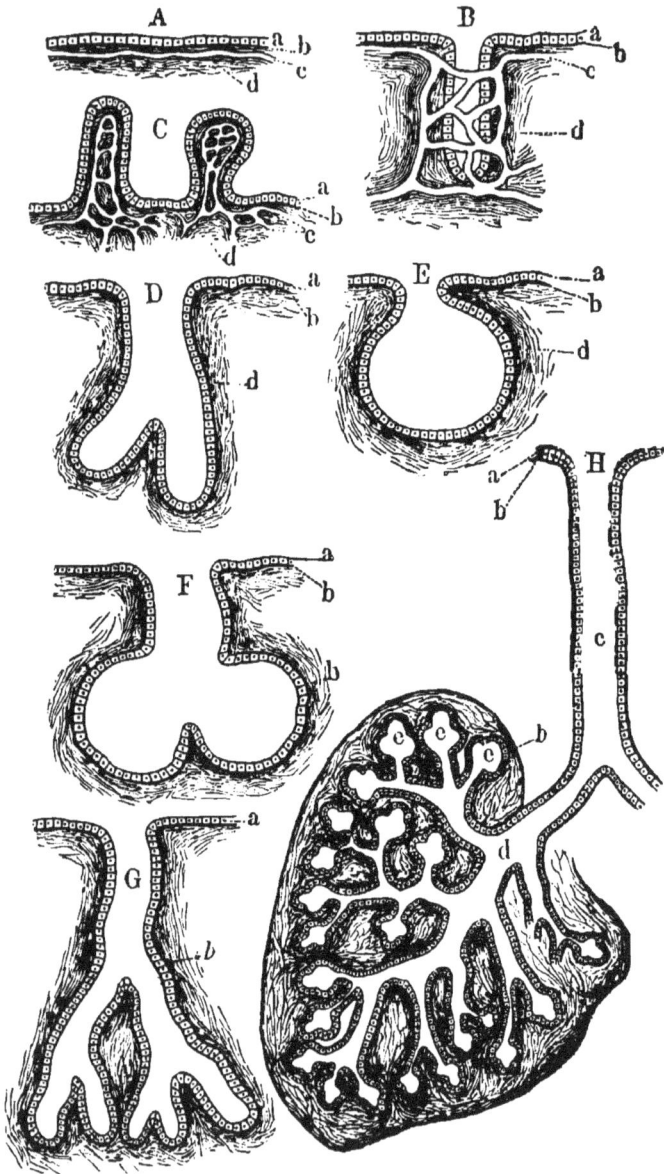

Fig. 40.—Forms of Glands. *A*, a simple secreting surface ; *a*, its epithelium; *b*, basement membrane ; *c*, capillaries ; *B*, a simple tubular gland ; *C*, a secreting surface increased by protrusions ; *E*, a simple racemose gland ; *D* and *G*, compound tubular glands; *F*, a compound racemose gland. In all but *A*, *B*, and *C* the capillaries are omitted for the sake of clearness. *H*, half of a highly developed racemose gland ; *c*, its main duct.

ways so that a wide surface can be packed in a small bulk, just as a Chinese paper lantern when shut up occupies much less space than when extended, although the actual area of the paper in it remains of the same extent. In a few cases the folding takes the form of protrusions into the cavity of the secreting organ, as indicated at *C*, Fig. 40, but much more commonly the surface extension is attained in another way, the supporting or *basement membrane*, covered by its epithelium, being pitted in or involuted as at *B*. Such a secreting organ is known as a true *gland*.

Forms of Glands.—In some cases the surface involutions are uniform in diameter, or nearly so, throughout (*B*, Fig. 40). Such glands are known as *tubular;* examples are found in the lining coat of the stomach (Fig. 48); also in the skin (Fig. 76), where they form the *sweat-glands*. In other cases the involution swells out at its deeper end and becomes more or less sacculated (*E*); such glands are named *racemose* or *acinous*. The small glands of the skin which form the oily matter poured out on the hairs (p. 303) are of this type. In both kinds the lining cells near the deeper end are commonly different in character from the rest; and around that part of the gland the finest and thinnest walled blood-vessels form a closer network. These deeper cells form the true secreting tissue of the gland, and the tube, lined with different cells, leading from the secreting recesses to the surface on which the secretion is poured out, and serving merely to drain it off, is known as the *duct* of the gland. When the duct is undivided the gland is *simple;* but when, as is more usual, it is branched and each branch

What is a tubular gland? Examples? A racemose gland? Example? Where do we find the closest network of blood-vessels in a gland? Which cells of a gland make its secretion? What is meant by the "duct" of a gland? What is a simple gland?

has a true secreting chamber at its end we get a compound gland, tubular (*G*) or racemose (*F*, *H*) as the case may be. In many cases the chief duct, in which the smaller ducts unite, is of considerable length, so that the secretion is poured out at some distance from the main mass of the gland.

A fully formed gland, *H*, is, therefore, a complex structure, consisting primarily of a duct, *c*, ductules, *dd*, and secreting recesses, *ee*. The ducts and ductules are lined with cells which are merely protective, and differ in character from the secreting cells which line the deepest parts. The cells lining the ultimate recesses differ in different glands, and produce different liquids; consequently, though all glands are built on much the same plan, they make very varied secretions, the nature of the secretion of any gland depending on the properties of its cells.

The Complexity of the Alimentary Canal.—We may now return to our immediate subject, the alimentary canal. This is not a simple tube, but presents several dilatations on its course; nor is it a comparatively straight tube, as diagrammatically represented in Fig. 1, but, being much longer than the regions of the body which it traverses, much of it is packed away by being coiled up in the abdominal cavity.

Subdivisions of the Alimentary Canal.—The mouth-opening leads into a chamber containing the teeth and tongue, and named the *mouth-chamber* or *buccal cavity*. This primary dilatation is separated by a constriction (*the*

A compound? How does it happen that the secretion is sometimes poured out at a distance from the main mass of the gland?

Describe a fully developed gland. How is it that glands make such different secretions? On what does the nature of the secretion of a gland depend?

How does the alimentary canal differ from a simple uniform tube? Why is a great part of it coiled?

Into what does the opening between the lips lead?

isthmus of the fauces) at the back of the mouth, from another, the *pharynx* or *throat chamber*, which narrows again at the top of the neck into the *gullet* or *œsophagus*, which runs as a comparatively narrow tube through the thorax, and then, passing through the diaphragm, dilates in the upper part of the abdominal cavity to form the *stomach* (see Fig. 1). Beyond the stomach the channel again narrows to form a long and greatly coiled tube, the *small intestine,* which terminates by opening into the *large intestine*, which, though shorter is wider, and ends by opening on the exterior.

The Mouth Cavity.—(Fig. 41) is bounded in front and on the sides by the lips and cheeks, below by the tongue, *k,* and above by the *palate,* which latter consists of an anterior part, *l,* supported by bone and called the *hard pal-*

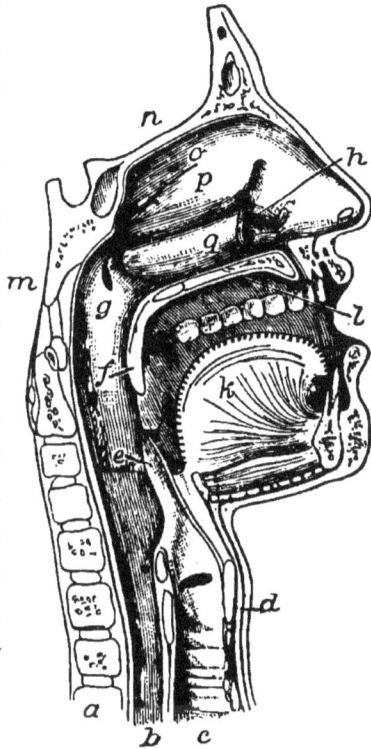

FIG. 41.—The mouth, nose and pharynx, with the commencement of the gullet and larynx, as exposed by a section, a little to the left of the median plane of the head. *a,* vertebral column ; *b,* gullet ; *c,* windpipe ; *d,* larynx; *e,* epiglottis; *f,* soft palate; *g,* opening of Eustachian tube ; *k,* tongue; *l,* hard palate; *m,* the sphenoid bone on the base of the skull; *n,* the fore part of the cranial cavity; *o, p, q,* the turbinate bones of the outer side of the left nostril chamber.

What is the isthmus of the fauces? Where does the gullet begin? Through what regions of the body does it pass? Where does the stomach lie? What part of the alimentary canal succeeds the stomach? Describe it briefly. How does it end? How does the large intestine differ from the small? How does it end?

What are the boundaries of the mouth cavity? Of what parts does the palate consist?

ate, and a posterior, *f*, containing no bone, and called the *soft palate*. The two can readily be distinguished by applying the tip of the tongue to the roof of the mouth and drawing it backwards. The hard palate forms the partition between the mouth and nose. The soft palate arches down at the back of the mouth, hanging like a curtain between it and the pharynx, as can be seen on holding the mouth open in front of a looking-glass. From the middle of its free border a conical process, the *uvula*, hangs down.

The Teeth.—Immediately within the cheeks and lips are two semicircles, formed by the borders of the upper and lower jaw-bones, which are covered by the *gums*, except at intervals along their edges where they contain sockets in which teeth are implanted. During life two sets of teeth are developed : the first or *milk set* appear soon after birth and are shed during childhood, when the second or *permanent set* appear.

The General Structure of a Tooth.—The teeth differ in minor points from one another, but in all, three parts are distinguishable ;* one, seen in the mouth, and called the *crown* of the tooth; a second, imbedded in the jaw-bone, and called the *root* or *fang;* and between the two, embraced by the edge of the gum, a narrowed portion, the *neck* or *cervix*. By differences in their forms and uses the teeth are divided into *incisors, canines, bicuspids,* and

How can we feel the difference between them? What cavities does the hard palate separate? Where does the soft palate lie? What is the uvula?

What do we find inside the lips? Where are the gums? What do the margins of the jaw bones contain ? What are the milk teeth? What the permanent?

What parts may we distinguish in every tooth? Into what groups are teeth divided? Why?

---

* A number of teeth can be readily obtained from a dentist, and will be found of great use in connection with this lesson.

molars, arranged in a definite order in each jaw. Beginning at the middle line we meet in each half of each jaw, successively, with two incisors, one canine, and two molars in the milk set; making twenty altogether in the two jaws. The teeth of the permanent set are thirty-two in number, eight in each half of each jaw, viz.—beginning at the middle line—two incisors, one canine, two bicuspids, and three molars. The bicuspids of the permanent set replace the molars of the milk set, while the permanent molars are new teeth added on as the jaw grows, and not substituting any of the milk teeth. The hindmost permanent molars are often called the *wisdom teeth*.

Characters of Individual Teeth.—The *incisors* or *cutting*

FIG. 42.   FIG. 43.   FIG. 44.   FIG. 45.

FIG. 42.—An incisor tooth.
FIG. 43 —A canine or eye tooth.
FIG. 44.—A bicuspid tooth seen from its outer side; the inner cusp is accordingly not visible.
FIG. 45.—A molar tooth.

teeth (Fig. 42) are adapted for cutting the food. Their crowns are chisel-shaped and have sharp horizontal cutting edges which become worn away by use, so that they are beveled off behind in the upper row and in the opposite

Enumerate the milk teeth in order. How many are there altogether? Number of permanent teeth? Enumerate in order. What permanent teeth replace the milk molars? What permanent teeth replace no milk teeth? Which are the wisdom teeth?
Describe an incisor tooth.

direction in the lower. Each has a single long fang. The *canines (dog teeth)* (Fig. 43) are somewhat larger than the incisors. Their crowns are thick and somewhat conical, having a central point or *cusp* on the cutting edge. In dogs and cats the canines are very long and pointed, and adapted for seizing and holding prey. The *bicuspids* or *premolars* (Fig. 44) are rather shorter than the canines and their crowns are cuboidal. Each has two cusps, an outer towards the cheek, and an inner on the side turned towards the interior of the mouth. The *molar teeth* or *grinders* (Fig. 45) have large crowns with broad surfaces, on which are four or five projecting tubercles which roughen them and make them better adapted to crush the food. Each has usually several fangs. The *milk teeth* differ only in subsidiary points from those of the same names in the permanent set.

The Structure of a Tooth.—If a tooth be broken open a cavity extending through both crown and fang will be found in it. This is filled during life with a soft pulp, containing blood-vessels and nerves, and is known as the "pulp cavity." The hard parts of the tooth disposed around the pulp cavity consist of three different tissues. Of these, one immediately surrounds the cavity and makes up most of the bulk of the tooth; it is *dentine* or ivory; covering the dentine on the crown is *enamel*, the hardest

A canine. Name animals with specially developed canines. For what do they use them? Give another name for a bicuspid tooth. Describe one.

Describe a molar tooth. What is the object of the projections on their crowns? How far do the milk teeth differ from the permanent in form?

What do we find on breaking open a tooth? What is it called? Why? What tissues form the hard parts of a tooth? Where does each lie?

tissue in the body,* and on the fang the *cement*, which is a thin layer of bone.

The pulp cavity opens below by a narrow aperture at the tip of the fang, or at the tip of each fang if the tooth has more than one. Through these openings its blood-vessels and nerves enter.

Hygiene of the Teeth.—The teeth should be thoroughly cleansed night and morning, by means of a tooth-brush dipped in tepid water ; once a day soap should be used, or a little very finely powdered chalk sprinkled on the brush. The weak alkali of the soap or chalk is useful. A large proportion of a tooth consists of carbonate of calcium, which readily dissolves in weak acids ; and decomposing food particles lodged between the teeth develop acids, which eat away the tooth slowly but surely. Hence all food particles should be carefully removed from between the teeth ; as this cannot always be effected completely it is important to brush the teeth with alkaline substances which will neutralize and render harmless any acid.†
Good manners forbid the public use of a tooth-pick, but on the earliest privacy after a meal a wooden or quill tooth-pick should be employed systematically and carefully to dislodge all food remnants which may have remained wedged between the teeth.

Where is the pulp cavity open? What things pass through the opening?
When and how should the teeth be cleansed? What substance forms a large part of the teeth? In what is this substance soluble? Why should food particles be carefully removed from between the teeth? Why are weak alkaline substances useful in cleaning the teeth?

---

* Enamel will strike fire with flint.
† Acid medicines should always be sucked up through a glass tube and swallowed with as little contact as possible with the teeth. After each dose the mouth should be thoroughly rinsed with water.

Once a slight cavity has been formed, the process of
decay is apt to go on very fast; first, because the ex-

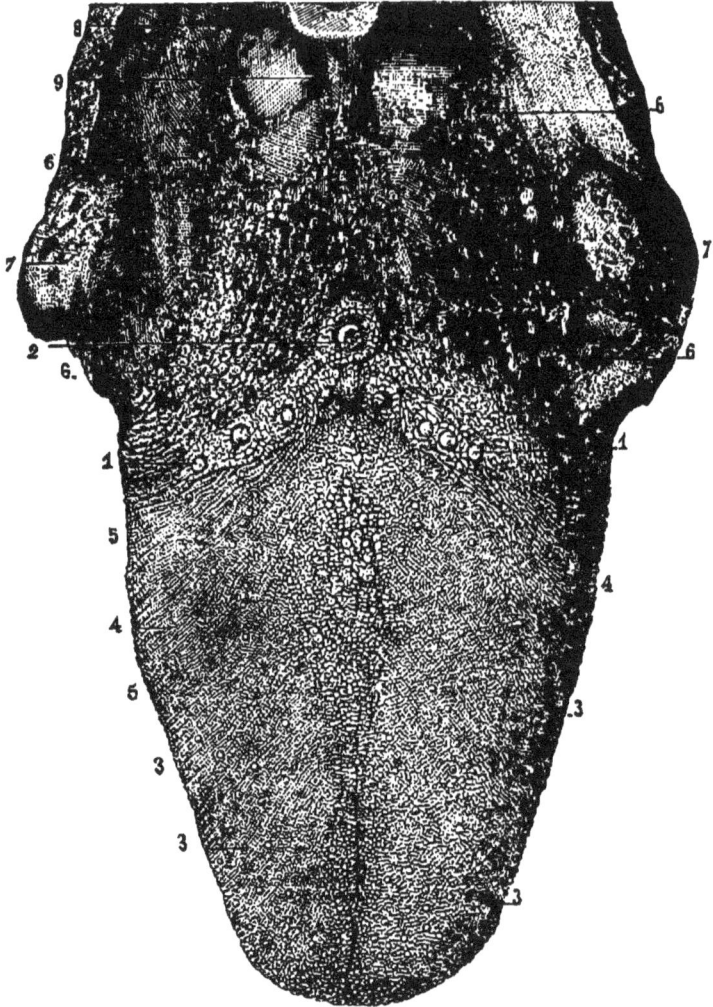

Fig. 46.- The upper surface of the tongue. 1, 2, circumvallate papillæ; **3**, fungiform papillæ; 4, filiform papillæ; 6, mucous glands.

posed deeper layer of the tooth is more easily dissolved
than its natural surface; and, second, because the little

pit forms a lodging-place for bits of food, which, in decomposing, form acids and hasten the corrosion. Small eroded cavities are very apt to be overlooked ; the teeth should, therefore, be thoroughly examined two or three times a year by a dentist.

The Tongue (Fig. 46) is a muscular and highly movable organ, covered by mucous membrane, and endowed not only with a delicate sense of touch, but with the sense of taste. Its root is attached to the hyoid bone (p. 38). The mucous membrane covering the upper surface of the tongue is roughened by numerous minute elevations or *papillæ*, of which there are three varieties. The *circum-vallate papillæ* (Fig. 46, 1 and 2) are the largest and fewest, and lie near the root of the tongue, arranged in the form of a V, with its open angle turned towards the lips. The *fungiform papillæ* are rounded masses attached by narrower stems. They are found all over the middle and fore part of the upper surface of the tongue, and in healthy persons are recognizable as red dots, more deeply colored than the rest of the mucous membrane. The *filiform papillæ* are pointed elevations scattered over the upper surface of the tongue, except near its root. They are on human tongues the smallest and most numerous.*

Why is decay of a tooth apt to go on fast once it has commenced ? Why should the teeth be examined from time to time by a dentist ?
Briefly describe the tongue. What sensations do we obtain through it? To what is its root attached? What are found on the mucous membrane of the upper surface of the tongue? Of how many varieties? Which are largest and fewest? Where are they found? How are they arranged? Describe the fungiform papillæ. Where are they found? What do they look like when we examine a person's tongue? Where are the filiform papillæ found? What is their form? What papillæ on the human tongue are smallest? Most numerous?

---

* The filiform papillæ are very large on the tongue of the cat, where they may readily be seen and felt. They are large in most of the carnivorous animals,

What a "Furred Tongue" Indicates.—In health the surface of the tongue is moist, covered by little "fur" and, in childhood, of a red color. In adult life the natural color of the tongue is less red, except around the edges and tip ; a bright red glistening tongue is then usually a symptom of disease. When the digestive organs are deranged the tongue is commonly covered with a thick yellowish coat, and there is frequently a "bad taste" in the mouth.* The whole alimentary mucous membrane is in close physiological connection ; and anything disordering the stomach is likely to produce a "furred tongue," which in most cases may be taken as indicating something wrong with the deeper parts of the digestive tract.

The Salivary Glands.—The saliva, which is poured into the mouth and moistens it, is secreted by three pairs of glands, the *parotid*, the *submaxillary*, and the *sublingual*. The parotid glands lie close in front of the ear ; each sends its secretion into the mouth by a duct, which opens inside the cheek opposite the second upper molar tooth. In the disease known as *mumps* † the parotid glands are inflamed and enlarged. The submaxillary glands lie between the halves of the lower jaw-bone, and their ducts open beneath

Describe the surface of a healthy tongue. How does the tongue of a healthy man differ in appearance from that of a healthy child? When is the tongue apt to be "coated"? What does a furred tongue usually indicate?

By what is the saliva secreted? Where does the parotid gland lie? Where does its duct open? What change occurs in the parotid glands during "mumps"? Where are the submaxillary glands? Where do their ducts open?

---

serving to scrape or lick clean bones, etc. Tamed tigers have been known to draw blood by licking the hand of their master.

* The fur of the tongue consists of some mucus, a few cells shed from its surface, and numerous vegetable microscopic organisms belonging to the group of *Bacteria*.

† Technically, *parotitis.*

the tongue. The sublingual glands lie beneath the floor of the mouth behind the submaxillary.

The Fauces is the name given to the passage which can be seen at the back of the mouth leading from it into the pharynx, below the soft palate.* It is bounded above by the soft palate and uvula, below by the root of the tongue, and on the sides by muscles, covered by mucous membrane, which reach from the soft palate to the tongue. The muscles cause elevations known as the *pillars of the fauces.* Each elevation divides near the tongue, and in the hollow between its divisions lies a *tonsil* (7, Fig. 46), a soft rounded body about the size of an almond, and containing numerous minute glands which form mucus.

Enlarged Tonsils.—The tonsils not unfrequently become enlarged during a cold or sore throat, and then pressing on the Eustachian tube (p. 373), which leads from the throat to the middle ear, keep it closed and cause temporary deafness. Sometimes the enlargement is permanent, and causes much annoyance. The tonsils can, however, be removed without much danger to life, and this is the treatment usually adopted in such cases.

The Pharynx or Throat Cavity (Fig. 41).—This portion of the alimentary canal may be described as a conical bag with its broad end turned towards the base of the skull and its other end turned downwards and narrowing into

Where do the sublingual glands lie?
What is meant by the fauces? How are they bounded? What are the pillars of the fauces? What is a tonsil?
Why is temporary deafness not uncommon when we have a sore throat? What is usually done when the tonsils are permanently enlarged?
Briefly describe the pharynx.

---

* Observe for yourself with the help of a looking-glass.

the gullet. Its front or ventral wall is imperfect, present-
ing apertures which lead into the nose, the mouth, and
(through the larynx and windpipe) into the lungs. Except
when food is being swallowed the soft palate hangs down
between the mouth and pharynx; during deglutition it is
raised into a horizontal position, and separates an upper or
respiratory portion of the pharynx from the rest. Through
this upper part air alone passes,* entering it from the posterior ends of the two nostril chambers, while through the
lower portion both food and air pass, one on its way to the
gullet, *b*, Fig. 41; the other through the larynx, *d*, to the
windpipe, *c*; when a morsel of food "goes the wrong way"
it takes the latter course. Opening into the upper portion
of the pharynx on each side is an Eustachian tube, *g*. At
the root of the tongue, over the opening of the larynx, is a
plate of cartilage, the *epiglottis*, *e*, which can be seen if the
mouth is widely opened and the back of the tongue pressed
down by some such thing as the handle of a spoon. During swallowing the epiglottis is pressed down like a lid over
the opening of the air-tube and helps to keep food from
entering it. The pharynx is lined by mucous membrane
and has muscles in its walls which, by their contractions,
drive the food on.

The Œsophagus or Gullet is a tube commencing at the

What apertures open into its ventral side? What is the usual
position of the soft palate? How is this position altered during
swallowing? What passes through the respiratory division of the
pharynx? What things pass through its lower division? What is
the destination of each? What is meant by saying a morsel has
"gone the wrong way"? Where do the Eustachian tubes open?
What is the epiglottis? How may it be seen? What is its use?

* During a severe attack of vomiting the soft palate often only acts imperfectly in closing the passage between gullet and nostrils; hence some of the
ejected matter not unfrequently is expelled through the nose.

lower termination of the pharynx and which, passing on through the neck and chest, ends below the diaphragm in the stomach. In the neck it lies close behind the windpipe.

The **Stomach** (Fig. 47) is a curved conical bag placed

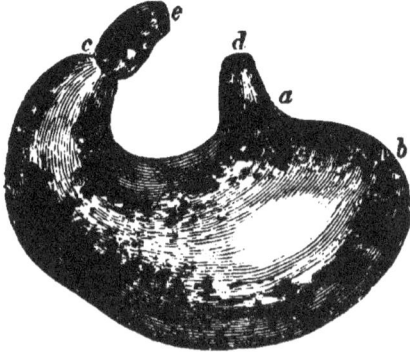

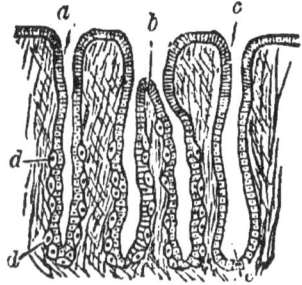

FIG. 47.   FIG. 48.

FIG. 47.—The stomach. *d,* lower end of the gullet ; *a,* position of the cardiac aperture : *b,* the fundus ; *c,* the pylorus ; *e,* the first part of the small intestine ; along *a, b, c,* the great curvature ; between the pylorus and *d,* the lesser curvature.

FIG. 48.—A thin section through the gastric mucous membrane, perpendicular to its surface, magnified about 25 diameters. *a,* a simple peptic gland; *b,* a compound peptic gland; *c,* a mucous gland.

transversely in the upper part of the abdominal cavity.* Its larger end is turned to the left and lies close beneath the diaphragm, and opening into its upper border, through the *cardiac orifice* at *a,* is the gullet, *d.* The narrower right end is continuous at *c* with the small intestine; the communication between the two is the *pyloric orifice.* The

Describe the gullet. Where does it lie in the neck?
What is the stomach? Which end of it is larger? Where does this end lie? What opens into it? What is the opening called? What is continuous with the small end of the stomach? What is the name of the aperture between the stomach and the small intestine?

---

* The general anatomical arrangement of the stomach, and its connections with the gullet and intestine, may be readily shown on the body of a puppy, kitten, or rat, which has been killed by placing it for five minutes in a small box containing also a sponge soaked with chloroform.

pyloric end of the stomach is separated from the diaphragm by the liver (see Fig. 4). When moderately distended the stomach is about twelve inches long, and about four inches across at its widest part, and would contain about three pints.

The Glands of the Stomach.—The mucous membrane lining the stomach is seen, when its surface is examined with a common magnifying glass, to be covered with shallow pits. A more powerful microscope shows on the bottom of each one of these pits the openings of several minute tubes, *the gastric glands*, which lie imbedded in the mucous membrane, packed closely, side by side (Fig. 48). These glands secrete the *gastric juice*.

The Muscular Coat of the Stomach lies outside the mucous membrane, and is made up (Fig. 34) of plain muscular tissue, whose fibres run in different directions. By its contractions it stirs up the food and mixes it with the gastric juice. Around the pyloric orifice of the stomach is a thick ring of muscle (the *pyloric sphincter*), which usually is contracted, closing the passage between the stomach and the commencement of the small intestine. During digestion in the stomach the pyloric sphincter relaxes from time to time, and allows food, more or less digested, to pass on into the intestine.

Palpitation of the Heart.—The cardiac end of the stomach lies close beneath the diaphragm, and the heart imme-

What lies between the right end of the stomach and the diaphragm? What is the size of the stomach?

What may be seen on examining the mucous membrane of the stomach with a hand lens? What does a more powerful magnifying instrument show? What is the function of the gastric glands?

Describe the muscular coat of the stomach. What is its function? What is the pyloric sphincter? Its function? What happens when the pyloric sphincter relaxes during gastric digestion?

diately above it. Over-distension of the stomach, due to indigestion or flatulency, may press up the diaphragm and interfere with the proper working of the thoracic organs, causing feelings of oppression in the chest, or palpitation of the heart.

The Small Intestine commences at the pylorus and ends, after many windings, in the large. It is about twenty feet (six meters) long and about two inches (five centimeters) wide at its gastric end, narrowing to about two thirds of that width at its lower portion. Externally there are no lines of subdivision on the small intestine, but anatomists arbitrarily describe it as consisting of three parts, the first twelve inches

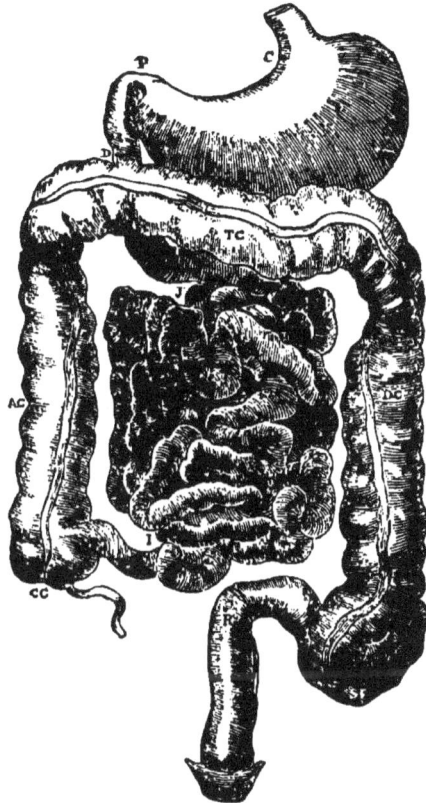

FIG. 49.—Diagram of abdominal part of alimentary canal. *C*, the cardiac, and *P*, the pyloric end of the stomach: *D*, the duodenum; *J*, *I*, the convolutions of the small intestine; *CC*, the cæcum with the vermiform appendix; *AC*, ascending, *TC*, transverse, and *DC*, descending colon; *R*, the rectum.

being the *duodenum*, the succeeding two fifths of the remainder the *jejunum*, and the rest the *ileum*.

Why is it that an over-distended stomach sometimes causes palpitation of the heart?

Where does the small intestine commence? Where does it end? Describe its length and diameter. Of what divisions do anatomists describe it as consisting?

The Mucous Coat of the Small Intestine.—This is pink, soft, and extremely vascular. It is throughout a great portion of the length of the tube raised up into permanent transverse folds in the form of crescentic ridges, each fold running transversely for a greater or less way round the intestine (Fig. 50). These folds are the *valvulæ conniventes*. They are first found about two inches from the pylorus, and are most thickly set and largest in the upper half of the jejunum, in the lower half of which they become gradually less conspicuous; they finally disappear altogether about the middle of the ileum. The folds of the mucous membrane

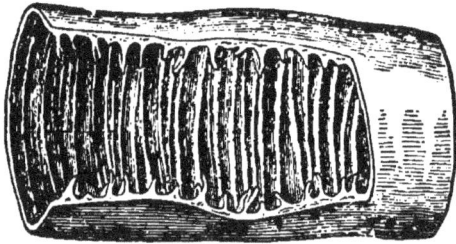

Fig. 50.—A portion of the small intestine opened to show the *valvulæ conniventes.*

serve to greatly increase its surface both for absorption and secretion, and they also delay the food in its passage; it collects in the hollows between them, and so is longer exposed to the action of the digestive liquids.

The Villi.—Examined closely with the eye or, better, with a hand lens, the mucous membrane of the small intestine is seen not to be smooth but shaggy, being covered everywhere (both over the valvulæ conniventes and between them) with closely packed minute elevations standing up somewhat like

Give the general characteristics of the mucous membrane of the small intestine. What are the valvulæ conniventes? Where do they commence? Where are they most developed? Where do they cease? What purposes do they subserve? What are the villi?

the "pile" on velvet and known as the *villi* (Fig. 51). In structure a villus is somewhat complex. Covering it is a single layer of cells, beneath which the villus may be regarded as made up of a framework of connective tissue supporting the more essential constituents. Near the surface is a network of plain muscular tissue. In the centre is an offshoot of the lymphatic or absorbent system, sometimes

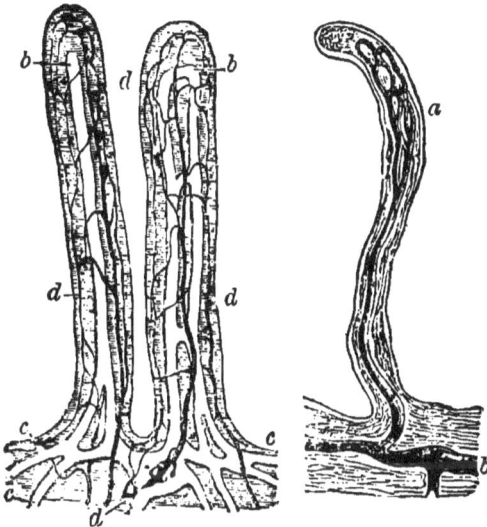

Fig. 51.—Villi of the small intestine; magnified about 80 diameters. In the left-hand figure the lacteals. *a*, *b*, *c*, are filled with white injection; *d*, blood-vessels. In the right-hand figure the lacteals alone are represented, filled with a dark injection. The epithelium covering the villi, and their muscular fibres are omitted.

in the form of a single vessel with a closed dilated end, and sometimes as a network formed by two main vessels with cross-branches. During digestion these lymphatics are filled with a milky white liquid absorbed from the intestines, and they are accordingly called the *lacteals*. They communicate with larger branches in the outer coats of the in-

Describe the structure of a villus. What is found in its lymphatics during digestion?

testine, and these end in trunks which join the main lymphatic system. Finally, in each villus, outside its lacteals and beneath its muscular layer, is a close network of bloodvessels.

The Glands of the Small Intestine.—Opening on the surface of the small intestine between the bases of the villi are small glands, the *crypts of Lieberkühn.* Each is a simple unbranched tube, lined by a single layer of cells.

The Muscular Coat of the Small Intestine lying outside the mucous coat, is composed of plain muscular tissue, disposed in two layers: an inner circular, and an outer longitudinal. By their combined and alternating contractions they slowly force the digesting food along the tube.

In the duodenum are found in addition minute glands, *the glands of Brunner,* which lie outside the mucous membrane, and send their ducts through it to open on its inner surface.

The Large Intestine (Fig. 49), forming the final portion of the alimentary canal, is about 5 feet (1.5 meters) long, and varies in diameter from 2½ to 1½ inches (6–4 centimeters). Anatomists describe it as consisting of the *cæcum* (CC) with its *vermiform appendix*, the *colon* (AC, TC, DC), and the *rectum* (R). The small intestine does not open into the end of the large but into its side, some distance from its closed upper end; the cæcum is that part of the large intestine which extends beyond the communication. From it projects the *vermiform appendix*, a narrow tube not thicker than a cedar pencil, and about 4 inches (10 centi-

Where do we find the crypts of Lieberkühn? Describe them. Where are the glands of Brunner?

Give the dimensions of the large intestine. Of what parts is it made up? How does the small intestine open into it? What is the cæcum? The vermiform appendix? Its size?

meters) long. The colon commences on the right side of the abdominal cavity where the small intestine communicates with the large, runs up for some way on that side (*ascending colon*), then crosses the middle line (*transverse colon*) below the stomach, and turns down (*descending colon*) on the left side, and there makes an S-shaped bend known as the *sigmoid flexure;* from this the *rectum* proceeds to the opening by which the intestine communicates with the exterior. The mucous coat of the large intestine possesses no villi nor valvulæ conniventes; it contains numerous closely set glands much like the crypts of Lieberkühn of the small intestine.

The Ileo-Colic Valve.—Where the small intestine joins the large there is a valve formed by two flaps of the mucous membrane sloping down into the colon, and so arranged as to allow matters to pass readily from the ileum into the large intestine, but not the other way.

The Liver.—Besides the secretions formed by the glands imbedded in its walls, the small intestine receives those of two large glands, the *liver* and *pancreas*, which lie in the abdominal cavity. The ducts of both open, by a common aperture, into the duodenum about 4 inches (10 centimeters) from the pylorus.

The *liver* is the largest gland in the body, weighing from 50 to 60 ounces (1400 to 1700 grams). It is

Describe the colon. What is the sigmoid flexure? What is the terminal portion of the alimentary canal named? How does the mucous lining of the large intestine differ from, and how does it resemble that of the small?

Where is the ileo-colic valve? How is it formed? What is its function?

What large glands pour their secretion into the small intestine? Where are they situated? Where do their ducts open?

What is the largest gland in the body? What is its weight? Where is it placed?

situated in the upper part of the abdominal cavity (*le, le'*, Fig. 4), rather more on the right than on the left side, immediately below the diaphragm. The liver is of dark reddish-brown color, and of soft friable texture. The vessels carrying blood to the liver (Fig. 52) are the *portal vein, Vp*, (p. 236) and the *hepatic artery;* both enter it at a groove on its under

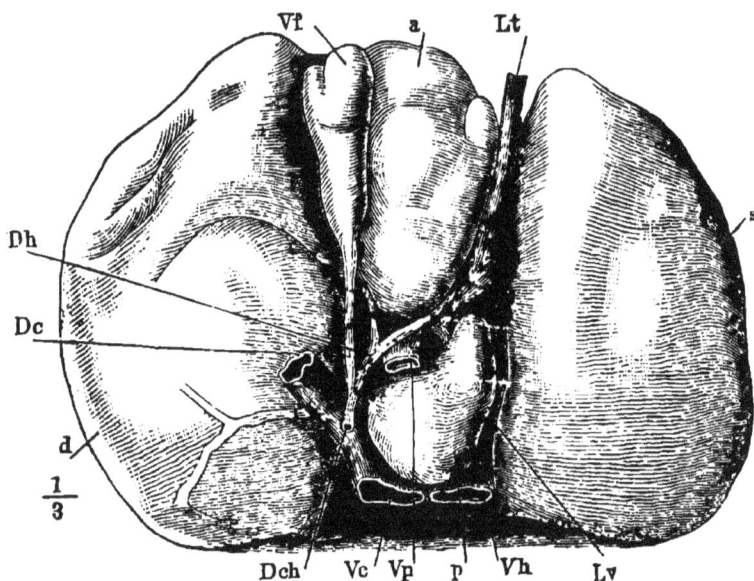

Fig. 52.—The under surface of the liver. *d*, right, and *s*, left lobe; *Vh*, hepatic vein; *Vp*, portal vein; *Vc*, vena cava inferior; *Dch*, common bile duct; *Dc*, cystic duct; *Dh*, hepatic duct; *Vf*, gall-bladder.

side, and there also a duct passes out from each half of the organ. The ducts unite to form the *hepatic duct, Dh,* which meets the *cystic duct, Dc,* proceeding from the *gall-bladder, Vf,* a pear-shaped sac in which the *bile* or *gall* formed by the liver accumulates when food is not being digested in the intestine. The *common bile duct, Dch,* formed

Describe the color and texture of the liver. What vessels bring blood to it? Describe the arrangement of its ducts. What is the gall-bladder? Where does the common bile duct open?

by the union of the hepatic and cystic ducts, opens into the duodenum.

The Functions of the Liver.—The size of the liver is related to the fact that the organ plays a double function; on the one hand it is a digestive gland secreting *bile;* on the other, its cells serve to store up, in the form of a kind of animal starch, called glycogen, excess of starchy or sugary food absorbed from the intestine during the digestion of a meal, and then to gradually dole this out to the blood for general use by the organs of the body until the next meal is eaten.

The Pancreas or Sweetbread * is a compound racemose gland. It is an elongated soft organ of a pinkish-yellow color, lying along the lower border of the stomach. Its right end is embraced by the duodenum which there makes a curve to the left. A duct traverses it and joins the common bile-duct close to its intestinal opening. The pancreas secretes a watery-looking liquid, much like saliva in appearance, which is of great importance in digestion.

With what fact is the large size of the liver connected? What are its functions?

To what group of glands does the pancreas belong? Describe its form and color. Where is it placed? What does its duct unite with? What does its secretion look like? Is it of much value?

---

* Butchers sell two kinds of sweetbread, known as the belly sweetbread and the neck or heart sweetbread. The former is the pancreas; the latter is the *thymus,* an organ of doubtful function, found only in young animals, and lying at the bottom of the neck and upper part of the chest in front of the windpipe.

CHAPTER XIII.

DIGESTION.

The Object of Digestion.—Some of the foodstuffs which we eat are already in solution and ready to soak at once into the lymphatics and blood-vessels of the alimentary canal; others, such as a lump of sugar, though not dissolved when put into the mouth, are readily soluble in the liquids found in the alimentary canal and need no further digestion. In the case of many most important foodstuffs, however, special chemical changes have to be brought about to make them soluble and capable of absorption. The different secretions poured into the alimentary tube act in various ways upon different foodstuffs, simply dissolving some and chemically changing others, until at last all are got into a condition in which they can be taken up into the lymph and blood-vessels for transference to distant parts of the body.

The Saliva.—The first solvent poured upon the food is the saliva, which, when it meets the food, is a mixture of pure saliva with the mucus secreted by the membrane lining the mouth. This *mixed saliva* is a colorless, cloudy, feebly alkaline liquid.

The Uses of Saliva are mainly physical and mechanical. It keeps the mouth moist and allows us to speak with com-

Explain the object of digestion.
What is the first digestive liquid which the food meets with?
How does it differ from pure saliva? Describe mixed saliva.

fort; most young orators know the distress occasioned by the suppression of the salivary secretion through nervousness, and the imperfect efficacy under such circumstances of the traditional glass of water placed beside public speakers. The saliva also enables us to swallow dry food; such a thing as a cracker when chewed would give rise merely to a heap of dust, impossible to swallow, were not the mouth cavity kept moist.* The saliva also dissolves such bodies as salt and sugar, when taken into the mouth in a solid form, and enables us to taste them; undissolved substances are not tasted, a fact which any one can verify for himself by wiping his tongue dry and placing a fragment of sugar upon it. No sweetness will be felt until a little moisture has exuded and dissolved part of the sugar.

Chemical Action of the Saliva.—In addition to such actions the saliva, however, exerts a chemical one on an important foodstuff. Starch (although it swells up greatly in hot water) is insoluble and could not be absorbed from the alimentary canal. The saliva has the power of turning starch into the readily soluble and absorbable grape sugar, the sugar of most fruits.† The starch is made to combine with the elements of water, and the final result is grape sugar.

Describe and illustrate the uses of saliva with reference (1) to speech, (2) to swallowing, (3) to dissolving some foods.
What foodstuff does saliva act upon chemically? What change is produced by its action?

---

* This fact used to be taken advantage of in the East Indian rice ordeal for the detection of criminals. The guilty person believing firmly that he cannot swallow the parched rice given him, and sure of detection, is apt to have his salivary glands paralyzed by fear, and so does actually become unable to swallow the rice; while in those with clear consciences the nervous system, acting normally, excites the usual reflex secretion, and the dry food causes no difficulty of deglutition.

† Grape sugar or *glucose* is now an extensively produced article of commerce, being made for this purpose by the prolonged action of dilute acids upon starchy substances.

$$C^{18}H^{30}O^{15} + 3H^2O = 3C^6H^{12}O^6$$

Starch.   Water.   Grape Sugar.

The Influence of Saliva in Promoting Digestion in the Stomach.—So far as chemical changes are concerned the saliva is but of secondary importance in digestion: its main use is to facilitate swallowing. It only changes starch into grape sugar (at least rapidly) when no acid is present, and food passes from the mouth to the stomach where it is mixed with the acid gastric juice, before the saliva has time to do much. Indirectly, however, the saliva promotes digestion in the stomach. Weak alkalies stimulate the gastric glands to pour forth more abundant secretion,* and the saliva, being alkaline, acts in this way. This is one reason why food should be well chewed before being swallowed; its taste, and the movements of the jaws, excite a more abundant salivary secretion, and this alkaline saliva, when swallowed, helps to stir the stomach up to work.

Swallowing or Deglutition.—A mouthful of solid food is broken up by the teeth and rolled about the mouth by the tongue until it is thoroughly mixed with saliva and made into a soft pasty mass. The muscles of the cheeks keep this from getting between them and the gums.† The mass is finally sent on from the mouth to the stomach by

What is the chief use of saliva? Under what circumstances does it change starch into sugar? In what portion of the digestive tract is this action of the saliva stopped? Why? How does the saliva promote digestion in the stomach? Why should food be thoroughly chewed before swallowing?

What is the technical term for swallowing? In how many stages does swallowing occur?

---

* Hence the efficacy of a little carbonate of soda or apollinaris water **taken** before meals, in some forms of dyspepsia.

† Persons with facial paralysis have from time to time to press out with the finger food which has collected outside the gums, where it can neither be chewed nor swallowed.

the process of *deglutition,* which occurs in three stages. The *first stage* includes the passage from the mouth into the pharynx. The food being collected into a heap on the tongue, the tip of that organ is placed against the front of the hard palate, and then the rest of the tongue is raised from before back, so as to compress the food mass between it and the palate and drive it through the fauces. This much of the act of swallowing is voluntary, or at least is under the control of the will, although it commonly takes place unconsciously. The *second stage of deglutition* is that in which the food passes through the pharynx; it is the most rapid part of its progress, since the pharynx has to be emptied quickly so as to clear the opening of the air-passages for breathing purposes. The food mass, passing back over the root of the tongue, pushes down the epiglottis ; at the same time the larynx (or voice-box at the top of the windpipe) is raised so as to meet the epiglottis, and thus the passage to the lungs is closed.* The soft palate is, at the same time, moved into a horizontal position, so as to separate the upper (or respiratory) portion of the pharynx, leading to the nose and the Eustachian tubes (see Fig. 41), from its lower portion, which ends inferiorly in the gullet.

Finally the isthmus of the fauces is closed as soon as the food has passed through, by the contraction of the muscles on its sides, and the elevation of the root of the tongue. All passages out of the pharynx except the gullet being thus blocked, when the pharyngeal muscles contract

Describe the first stage. What is the second stage? Which stage is most rapid? Why? How is the passage to the lungs closed while food is passing through the pharynx? How is the passage to the nose blocked? Describe the processes of the second stage of deglutition.

---

* The raising of the larynx during swallowing can be readily felt by placing the finger on its large cartilage forming "Adam's apple" in the neck.

the food can only be squeezed into the œsophagus. The muscular movements concerned in this part of deglutition are all excited without the intervention of the will; food touching the mucous membrane of the pharynx produces quite involuntarily the proper action of the swallowing muscles.* Indeed, many persons after having got the mouth completely empty cannot perform the movements of the second stage of deglutition at all. On account of the involuntary nature of the contractions of the pharynx any food which has once entered it must be swallowed; the isthmus of the fauces forms a sort of Rubicon; food that has entered the pharynx must be swallowed, even although the swallower learned immediately that he was taking poison. The *third stage of deglutition* is that in which the food is passing along the gullet, and is comparatively slow. Even liquid substances do not fall or flow down this tube, but have their passage controlled by its muscular coats, which grip the successive portions swallowed and pass them on. Hence the possibility of performing the apparently wonderful feat of drinking a glass of water while standing upon the head, often exhibited by jugglers; people forgetting that one sees the same thing done every day by horses and other animals which drink with the pharyngeal end of the gullet lower than the stomach.

The Gastric Juice.—The food having entered the stomach is exposed to the action of the gastric juice, which is a thin colorless or pale yellow liquid of a strongly acid reaction. It contains, beside water and some salts and mu-

How are the movements of the second stage of deglutition excited? What is the third stage of deglutition? Is it fast or slow? How is it that jugglers can drink while standing on the head? Describe the gastric juice.

---

* The process is what is known as a reflex action. See Chap. XX.

cus, *free hydrochloric acid* (about .02 per cent), and a substance called *pepsin*, which in acid liquids has the power of converting ordinary albumens into closely allied bodies called *peptones.* It also dissolves solid proteids, changing them at the same time into peptones.

Peptones.—Ordinary albumens are typical examples of what are called "colloids;" that is to say, substances which do not readily pass through moist animal membranes; peptones are kinds of albumen which do readily pass through such membranes, and are, therefore, capable of absorption from the alimentary canal. (See *Dialysis,* p. 212.)

Gastric Digestion.—In the stomach the onward progress of the food is stayed for some time. The pyloric sphincter remaining contracted closes the aperture leading into the intestine, and the irregularly disposed muscular layers of the stomach keep its semi-liquid contents in constant movement, by which all portions are thoroughly mixed with the secretion of its glands. In the stomach part of the albumen of the food is dissolved and turned into peptones. Certain mineral salts (as phosphate of lime, of which there is always some in bread), which are insoluble in water but soluble in dilute acids, are also dissolved in the stomach. On the other hand, the gastric juice has no action upon starch, nor does it digest oily substances. By the solution of the white fibrous connective tissues the disintegration of animal foods, commenced by the teeth, is

Name its more important constituents. What powers does pepsin possess?

What are colloids? Give examples. How do peptones differ from other albumens?

Does food pass on immediately from the stomach to the intestine? How is it kept back? How is it mixed with the gastric juice? What happens to albuminous foods in the stomach? Name another substance dissolved in the stomach. Name foodstuffs which are not changed in the stomach. How are animal foods broken up in the stomach?

carried much further in the stomach; and the food-mass, mixed with much gastric secretion, becomes reduced to the consistency of a thick soup, usually of a grayish color. In this state it is called *chyme.*

The Chyme contains, after an ordinary meal, a considerable quantity of peptones, which are in great part gradually absorbed into the blood and lymphatic vessels of the gastric mucous membrane and carried off, along with other dissolved and dialyzable bodies—for example, salts and sugar. After the food has remained in the stomach some time (one and a half to two hours) the chyme begins to be passed on into the intestine in successive portions. The pyloric sphincter relaxes at intervals, and the rest of the stomach, contracting at the same moment, injects a quantity of chyme into the duodenum; this is repeated frequently, the larger undigested fragments being at first unable to pass the orifice. At the end of three or four hours after an ordinary meal the stomach is quite emptied, the pyloric sphincter finally relaxing to such an extent as to allow any larger indigestible masses, which the gastric juice has not broken down, to be squeezed into the intestine.*

The Chyle.—When the chyme passes into the duodenum it finds preparation made for it. The pancreas commences to secrete as soon as food enters the stomach; hence a quantity of its secretion is already accumulated in the intestine when the chyme enters. The gall-bladder is distended

What is chyme? What things would be found in chyme after an ordinary meal? When does chyme begin to be sent on to the intestine? How? How soon is the stomach completely emptied after a meal? What has accumulated in the small intestine when the chyme reaches it ?

---

* Several of the above facts were first observed on a Canadian trapper. Alexis St. Martin, who as a result of a gunshot wound had a permanent opening from the surface of the abdomen to the interior of the stomach.

with bile, secreted since the last meal; the acid chyme stimulating the duodenal mucous membrane causes, through the nervous system, a contraction of the muscular coat of the gall-bladder, and so a gush of bile is poured out on the chyme. From this time on both liver and pancreas continue secreting actively for some hours, and pour their products into the intestine. The glands of Brunner and the crypts of Lieberkühn are also set at work. All of these secretions are alkaline, and they suffice very soon to more than neutralize the acidity of the gastric juice, and so to convert the acid *chyme* into alkaline *chyle*, which, as found in the intestine after an ordinary meal, contains a great variety of things: water, partly swallowed and partly derived from the salivary and other secretions; some undigested albumens; some unchanged starch; oils from the fats eaten; peptones formed in the stomach but not yet absorbed; salines and sugar, which have also escaped complete absorption in the stomach; indigestible substances taken with the food; all mixed with the secretions of the alimentary canal.

The **Pancreatic Secretion** is clear, watery, alkaline, and much like saliva in appearance. The Germans call the pancreas the "abdominal salivary gland." In digestive properties, however, the pancreatic secretion is far more important than the saliva, acting not only on starch, but on albumens and fats. On starch it acts like the saliva, but more energetically. It produces changes in albumens similar to those effected in the stomach, but by the agency of a

How is an outpouring of bile on the chyme brought about? Do liver and pancreas cease secreting when the chyme enters the intestine? What other glands are set to work? How is the acidity of the chyme overcome? What is chyle? What does it usually contain?

Describe the pancreatic secretion. What foodstuffs does it act upon? Describe its action on starch. How does it change albumens?

different substance, *trypsin,* which differs from pepsin in acting in an alkaline instead of in an acid medium. On fats it has a double action. To a certain extent it breaks them up into fatty acids and glycerine.* The fatty acid then combines with some of the alkali present to make a *soap,* which being soluble in water is capable of absorption.† Glycerine also is soluble in water and capable of absorption. The greater part of the fats is not, however, so broken up, but simply mechanically separated into little droplets which remain suspended in the chyle and give it a whitish color; just as cream-drops are suspended in milk, or olive oil in mayonnaise sauce. If oil be shaken up with water, the two cannot be got to mix; immediately the shaking ceases the oil floats up to the top; but if some raw egg be added a creamy mixture is readily formed in which the oil remains for a long time evenly suspended in the watery menstruum. The reason of this is that each oil-droplet becomes surrounded by a delicate pellicle of albumen, and is thus prevented from fusing with its neighbors to make large drops which would soon float to the top. Such a mixture is called an *emulsion,* and the albumen of the pancreatic secretion emulsifies the oils in the chyle, which becomes white (for the same reason as milk is that

How does trypsin differ from pepsin ? How does pancreatic secretion break up some fats ? What digestive end is thus attained ? How is most of the fat eaten acted upon by the pancreatic secretion ? Why is the chyle white ? How may we mix oil with water ? Explain the process. What is an emulsion ? What emulsifies the oily matters of the chyle ?

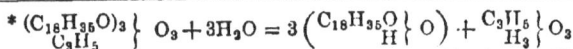

---

* $(C_{18}H_{35}O)_3 \atop C_3H_5 \} \ O_3 + 3H_2O = 3 \left({C_{18}H_{35}O \atop H} \} \ O \right) + {C_3H_5 \atop H_3} \} \ O_3$

　1 Stearin　+　3 Water　=　3 Stearic acid　+　1 Glycerine.

† Ordinary soap is a compound of a fatty acid with soda, colored and scented by the addition of various substances. Soft soap is a compound of a fatty acid with potash. Both dissolve in water, which the fats from which they are made will not do.

color) because the innumerable tiny oil-drops floating in it reflect all the light which falls on its surface.

The Bile.—Human bile when quite fresh is a golden brown liquid. It is alkaline, and besides coloring matters, mineral salts and water, contains the sodium salts of two nitrogenized acids, *taurocholic* and *glycocholic*, the former predominating in human bile.

The Uses of Bile.—Bile has no digestive action upon starch or albumens. It does not break up fats, but to a limited extent emulsifies them when shaken up with them outside the body, though far less perfectly than the pancreatic secretion. It is even doubtful if this action is exerted in the intestines at all. In many animals, as in man, the bile and pancreatic ducts open together into the duodenum, so that on killing the creature during digestion and finding emulsified fats in the chyle it is impossible to say whether or no the bile had a share in the work. In rabbits, however, the pancreatic duct opens into the intestine about a foot farther from the stomach than the bile-duct; and it is found that if a rabbit be killed after being fed with fatty food, no milky chyle is found above the point where the pancreatic duct opens. In the rabbit, therefore, the bile alone does not emulsify fats; and since bile is much the same in rabbits and other mammals, it probably does not emulsify fats by its reaction in any mammals. The inertness of bile in digestion has caused it to be doubted whether it is of any use, and whether it should not be regarded merely as an excretion, poured into the aliment-

Describe fresh human bile. What is its reaction? Name its chief constituents.

Name foods on which bile has no influence? How does it act upon fats when shaken with them? Give a reason for doubting if it emulsifies fats in the intestine.

ary canal to be got rid of. But there are many reasons against such a view. In the first place, the entry of the bile into the upper end of the small intestine, where it has to traverse a course of more than twenty feet before getting out of the body, makes it probable that bile has some function to fulfill in the intestine. One use is no doubt to assist by its alkalinity in overcoming the acidity of the chyme, and so to allow the pancreatic secretion to act upon proteids. Constipation is also apt to occur in cases where the bile-duct is temporarily stopped, so that the bile probably helps to excite the contractions of the muscular coats of the intestines; and it is said that when the bile secretion is deficient putrefactive changes are extremely apt to occur in the intestinal contents. Apart from such secondary actions, however, the bile probably has some influence in promoting the absorption of fats. If one end of a very narrow glass tube moistened with water be dipped in oil the latter will not rise in it, or but a short way; but if the tube be moistened with bile instead of water the oil will ascend higher. Again, oil passes through a plug of porous clay kept moist with bile, under a much lower pressure than through one wet with water. Hence bile by moistening the cells lining the intestine may facilitate the passage into the villi of oily substances. At any rate, experiment shows that if the bile be prevented from entering the intestine of a dog the animal eats an enormous amount of food compared with that amount which it needed previously; and that of this food

Give reasons for believing that bile is not a mere excretion. How does bile aid the digestive power of the pancreas? Point out other uses of bile. Describe experiments which tend to prove that bile helps in promoting the absorption of fatty matters from the intestine.

a great proportion of the fatty part passes out of the alimentary canal unabsorbed. There is no doubt therefore that the bile somehow aids in the absorption of fats.

Intestinal Juice or **the Succus Entericus** consists of the mixed secretions of the glands of Brunner and the crypts of Lieberkühn. It is very difficult to obtain it pure, and hence its digestive action is but imperfectly known. It is alkaline, and so helps to overcome the acidity of the chyme and allow the trypsin of the pancreas to act on albumens, and seems capable itself of dissolving some kinds of albumens, and turning them into peptones.

Intestinal Digestion.—Having considered separately the digestive actions of the different secretions poured into the small intestine, we may now consider their combined action. The acid chyme entering the duodenum from the stomach is more than neutralized by the alkaline secretions which it meets in the small intestine; it is made alkaline. This alkalinity allows the pancreatic secretion to finish the solution and transformation into peptone of proteids which have escaped conversion in the stomach. The pancreatic secretion also continues that conversion of insoluble starch into soluble and absorbable grape sugar, which had commenced in the mouth but was checked in the stomach. The bile and pancreatic secretion together emulsify the fats, with which they are thoroughly mixed by the contractions of the muscular coat of the intestine; they get them into a state of very fine division in the form of microscopic droplets, which are taken up by the cells lining the intestine. To a certain extent the fats are also saponified. The

What does the succus entericus consist of? Why is its digestive action but little known? Point out some of its uses.
Describe the process of intestinal digestion.

result of all these processes is a thin, milky looking alkaline liquid called *chyle*.

Indigestible Substances. — With every meal several things are eaten which are not digestible at all. Among them is *elastic tissue*, forming a part of the connective tissue of all animal foods, and *cellulose*, which is the chief constituent of the cases which envelope the cells of plants. The mucus secreted by the membrane lining the alimentary tract also contains an indigestible substance, *mucin*. These three materials, together with some water, some undigested foodstuffs, and some excretory substances found in the various secretions poured into the alimentary canal, form a residue which collects in the lower end of the large intestine, and is from time to time expelled from the body.

Dyspepsia is the common name of a variety of diseased conditions attended with loss of appetite or troublesome digestion. Being often unattended with acute pain, and if it kills at all doing so very slowly, it is pre-eminently suited for treatment by domestic quackery. In reality, however, the immediate cause of the symptoms, and the treatment called for, may vary widely; and the detection of the cause and the choice of the proper remedial agents often call for more than ordinary medical skill. A few of the more common forms of dyspepsia may be mentioned here, with their proximate causes, not in order to enable people to undertake the rash experiment of dosing them-

Name some indigestible substances eaten in every ordinary meal. Point out the source of each. What indigestible substance is added in the alimentary canal? What substances are found in the lower end of the large intestine?

What is meant by dyspepsia? Why is it not a wise thing for people to try to treat it themselves without skilled advice?

selves, but to show how wide a chance there is for any unskilled treatment to miss its end and do more harm than good.

Appetite is primarily due to a condition of the mucous membrane of the stomach, which in health comes on after a short fast and stimulates its sensory nerves ; and loss of appetite may be due to any of several causes. The stomach may be apathetic and lack its normal sensibility so that the empty condition does not act, as it normally does, as a sufficient excitant. When food is taken it is a further stimulus and may be enough ; in such cases "appetite comes with eating." A little quinine or tincture of gentian before a meal is often useful to patients of this class. On the other hand, the stomach may be too sensitive, and a voracious appetite be felt before a meal, which is replaced by nausea, or even vomiting, as soon as a few mouthfuls have been swallowed; the extra stimulus of the food then over-stimulates the too irritable stomach, just as a draught of mustard and warm water will a healthy one. The proper treatment in such cases is a soothing one.* In states of general debility, when the stomach is too feeble to secrete under any stimulation, the administration of weak acids and artificially prepared pepsin is needed, so as to supply gastric juice from outside until the improved

Describe the symptoms of some chief forms of dyspepsia.

---

* When food is taken it ought to stimulate the sensory gastric nerves, so as to excite the reflex centres for the secretory nerves and for the dilatation of the blood-vessels of the organ; if it does not, the gastric juice will be imperfectly secreted. In such cases one may stimulate the secretory nerves by weak alkalies (p. 174), as apollinaris water or a little carbonate of soda, before meals; or give drugs, as strychnine, which increase the irritability of reflex nerve-centres. The vascular dilatation may be helped by warm drinks, and this is probably the *rationale* of the glass of hot water after eating which has recently been in vogue; the usual cup of hot coffee after dinner is a more agreeable form of the same aid to digestion.

digestion strengthens the stomach up to the point of being able to do its own work.

Enough has probably been said to show that dyspepsia is not a disease, but a symptom accompanying many diseased conditions, requiring special knowledge for their treatment. From its nature—depriving the body of its proper nourishment—it tends to intensify itself, and so should never be neglected; a stitch in time saves nine.

Absorption from the Alimentary Canal.—Through its whole extent the mucous membrane lining the digestive tube is traversed by very closely packed tubes of two kinds, the blood and lymph vessels. Matters ready for absorption pass through or between the cells covering the surface of the mucous membrane, and then through the very thin walls of the smallest blood and lymph vessels; and by these vessels are conveyed to larger channels with thicker walls, which all ultimately lead to the heart. From the heart the digested and absorbed food is distributed to every organ of the body.

Absorption from the Mouth, Pharynx, and Gullet is but slight. Some water, some common salt, some sugar, and some grape sugar (made from starch by the action of saliva) are no doubt taken up during the processes of chewing and swallowing. But the time which elapses between taking a mouthful of food and its transference to the stomach is usually too short to allow the occurrence of any considerable absorption.

Why should dyspepsia never be neglected?

What tubes are found in the mucous membrane of the alimentary canal? How do dissolved foods enter them? Where are the absorbed matters carried? To what parts are they finally distributed?

What foodstuffs are partly absorbed in mouth, pharynx, and gullet? Why does not any great amount of absorption take place in those parts?

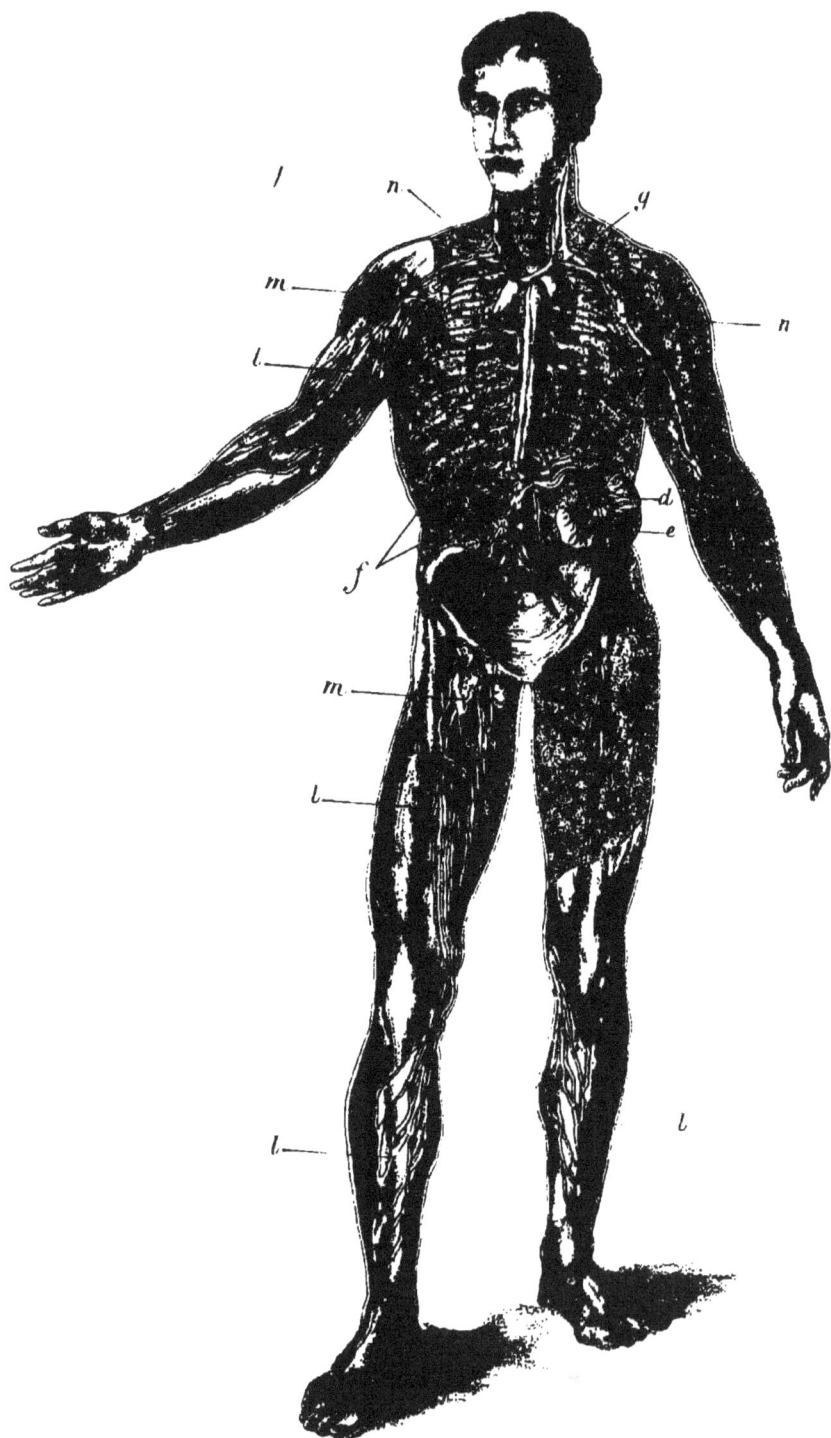

PLATE III.—A GENERAL VIEW OF THE LYMPHATICS OR ABSORBENTS. That portion of them known as the lacteals is seen at *d*, passing from the small intestine *e* to the thoracic duct *f*.

EXPLANATION OF PLATE III.

A General View of the Lymphatic or Absorbent System of Vessels.

e, A portion of the small intestine from which lacteals or chyle-conveying vessels, *d*, proceed, their origin within the villi may be seen magnified in fig. 51; *f*, the duct called thoracic, into which the lacteals open. This duct passes up the back of the chest, and opens into the great veins at *g*, on the left side of the neck: here the chyle mingles with the venous blood. In the right upper, and lower limbs the superficial lymphatic vessels *l l l l*, which lie beneath the skin, are represented. In the left upper and lower limbs the deep lymphatic vessels which accompany the deep blood-vessels are shown. The lymphatic vessels of the lower limbs join the thoracic duct at the spot where the lacteals open into it: those from the left upper limb and from the left side of the head and neck open into that duct at the root of the neck. The lymphatics from the right upper limb and from the right side of the head and neck join the great veins at *n*. *m m*, enlargements called lymphatic glands, situated in the course of the lymphatic vessels. These vessels convey a fluid called lymph, which mingles with the blood in the great veins. A fuller account of the lymphatic vessels in general, as distinguished from that section of them known as the lacteals, will be found on p. 214.

Absorption from the Stomach is more important. Food stays there a considerable time, and a good deal of the substances mentioned above as being absorbed to a slight degree on their way to the stomach, are taken up to a much greater extent by the mucous membrane of the stomach itself and passed on into the general blood current. In addition, a large proportion of albuminous food is turned in the stomach into peptones, which can be and are readily absorbed by the vessels of the gastric mucous membrane.

Absorption from the Small Intestine is by far the most important in bringing nutritive matters into the body. The stomach is an organ rather of digestion than absorption; the small intestine, on the other hand, is specially constructed to absorb. Its *valvulæ conniventes* delay the progress of the food mass, which stagnates in the hollows between them; and its innumerable villi, with their blood-vessels and lymphatics (p. 167), reach out, like so many rootlets, into the chyle and take it up.

The sugars reaching the small intestine or formed in it are absorbed mainly by the blood-vessels and carried to the liver, where they are turned into *glycogen* (p. 171), which is heaped up in the liver during digestion, and slowly given out to the blood, as its sugar is used up gradually before the next meal. The peptones passed into the intestine from the stomach, or formed in it by the action of the

Why does more absorption take place from the stomach? Name things absorbed from both mouth and stomach? What food matters are first absorbed from the stomach?

Where does the most important food absorption occur? What structural peculiarities of the small intestine peculiarly fit it for absorbing?

What vessels absorb sugars in the small intestine? To what organ are these sugars conveyed? What there becomes of them?

pancreatic secretion, are partly taken up by its lymphatics and partly by its blood-vessels. The emulsified fats mainly pass into the lymphatics of the villi, and are carried off by them.

The Lacteals.—The innumerable tiny fat drops drained off by the intestinal lymphatics or *lacteals* after an ordinary meal make their contents look white and milky, hence the name.* During fasting the lymphatics of the small intestine, like those in other parts of the body (see Chap. XIII.) convey a clear colorless liquid.

Absorption from the Large Intestine.—In the duodenum the bulk of food entering from the stomach is increased by the bile and pancreatic secretions poured out on it. Thenceforth absorption overbalances excretion, and the food-mass becomes less and less in bulk to the lower end of the ileum. The contractions of the small intestine drive on its continually diminishing contents, until they reach the ileo-colic valve, through which they are ultimately pressed. When the mass enters the large intestine its nutritive portions have been almost entirely absorbed, and it consists chiefly of some water, with the indigestible portions of the food and of the secretions of the alimentary canal. It contains cellulose, elastic tissue, mucin, and somewhat altered bile pigments; commonly some fat if a large quantity has been eaten; and some starch, if raw veg-

How are emulsified fats carried off?
What are the lacteals? Why so called? Under what conditions do the lacteals not contain milky looking chyle?
In what part of the alimentary canal does absorption more than balance the amount of liquid poured out on the food?
What are the constituents of the mass passing from the small into the large intestine? What changes does this mass undergo as it passes along the large intestine?

---

* From Latin, *lac*, milk.

etables have formed part of the diet. In its progress through the large intestine the food-mass loses still more water, and the digestion of starch and the absorption of fats is continued. Finally the residue, with some excretory matters added to it in the large intestine, is expelled from the body.

Pure Alcohol placed on the skin evaporates very rapidly, and in so doing abstracts heat, producing a sensation of coolness. This is succeeded by a feeling of warmth in the part; which also becomes red from temporary paralysis of its blood-vessels, causing them to dilate. If the evaporation be prevented, as by putting a little alcohol on the skin and covering it with a thimble, the alcohol acts as an irritant; it causes smarting, and finally sets up inflammation.

On the mucous membranes of the alimentary canal alcohol acts much as on the skin, but its irritant effect is more marked. Placed on the tongue it causes a feeling of coolness, followed by a hot biting sensation, and a red congested condition of the area with which it came in contact. Introduced into the stomach, where it cannot readily evaporate, strong alcohol causes congestion and inflammation varying in intensity with its amount. If the dose is large it sometimes causes death almost instantly, because the powerfully irritated sensory nerves of the gastric mucous membranes reflexly excite a nerve-centre which stops the heart's beat.

Diluted Alcohol.—Alcohol in such proportion as it exists in most alcoholic drinks is absorbed and carried in the blood

How does strong alcohol affect the skin? The alimentary mucous membranes? How may a large dose cause sudden death?

and lymph through the body, and if steadily taken day after day acts upon and alters for the worse nearly every important organ. The organ first or most seriously attacked varies with the form in which the alcohol is taken, with the amount consumed daily, and with the constitution of the individual. Probably no one individual ever suffered from all the diseased states produced by alcohol; but all habitual drinkers sooner or later experience one or more of them. The diseases produced by alcohol after absorption into the blood come on so gradually (except in the case of obvious drunkards) that the victim rarely perceives them until serious if not irremediable damage has been done: indeed, physicians have only recently come to clearly recognize that men who in common phrase " were never in their lives under the influence of liquor" may nevertheless be drinking enough to do them grave injury.

When constantly irritated by the direct action of alcoholic drinks, **the Stomach** gradually undergoes lasting structural changes. Its vessels remain dilated and congested, its connective-tissue becomes excessive, its power of secreting gastric juice diminished, and its mucous secretion abnormally abundant.

A vast number of persons suffer from *alcoholic dyspepsia* without knowing its cause; people who were never drunk in their lives, and consider themselves very temperate. "The symptoms vary, but when slight are something like these: A man (or woman) complains of slight loss of appetite, especially in the morning for breakfast; feels languid either on rising or early in the day; retches

How does diluted alcohol influence the stomach? Describe alcoholic dyspepsia. Its cure.

a little in the morning, and perhaps brings up a little phlegm only, or may actually vomit; or may be able to take breakfast, but feels sick after it. Towards the middle of the morning he is heavy and languid, perhaps, and does not feel easy until he has had a glass of sherry or some spirits, then gets on pretty well, and can eat lunch or dinner. Or if worse, the appetite for both is defective, and there is undue weight or discomfort after meals. . . . All these symptoms *may* be due to other causes, but when taken together they are by far most commonly due to alcohol." Abstinence from alcohol, the cause of the trouble, is the true remedy.

Effects of Alcohol upon the Liver.—Another serious bodily deterioration produced by alcohol is *fibrous degeneration :* by this is meant an excessive growth of the connective-tissue, which as we have seen (p. 36) pervades the organs of the body as a fine supporting skeleton for the more essential cells. Alcohol drinking causes this tissue to develop to such an extent as to crush and destroy the cells, especially in the liver and kidneys, which it should protect. So far as the liver is concerned, the result is a shrunken, rough mass (*hob-nailed liver*, or *gin-drinker's liver*), with hardly any liver-cells left in it. This not only prevents the proper manufacture of bile and glycogen (p. 171), but the contracted liver presses on the branches of the portal vein within it (p. 234) so as to impede the drainage of blood from the organs in the abdomen. As a consequence, an excess of the watery part of the blood oozes into the peritoneal cavity and accumulates, causing abdominal dropsy (*ascites*).

Describe the effects of alcohol upon the liver.

The Liver also suffers fatty degeneration, and is one of the organs most often and earliest attacked. This we might expect, as alcohol absorbed from the stomach is carried directly to the liver by the portal vein (p. 236). Alcohol also increases the breaking down of proteid matter in the body; the liver has much to do in preparing this broken-down proteid for removal by the kidneys, and so gets over-worked.

APPENDIX TO CHAPTER XIII.

The digestion and absorption of food are such fundamental facts in physiology that a thoroughly intelligent comprehension of them is of great importance; at the same time they are so largely merely chemico-physical phenomena that they are readily illustrated by a few simple experiments. These described below take but little time and cost but little money, while they cannot fail to be of value not merely in interesting a class, but in giving its members a much better idea of the way in which food is digested than they can get from merely reading a book.

1. Anatomy of the Alimentary Canal.—Kill a rat by chloroform or drowning. Dissect away the skin from the whole ventral aspect of the body.

Note in the neck region the large *salivary glands* which meet in the middle line: the posterior gland, close to the middle line, rounded and compact, is the *submaxillary;* on raising it, its *duct* will be seen passing forwards to the mouth, into which it may be followed by separating the halves of the lower jaw.

The large gland, composed of several loosely united lobes, and reaching from the neighborhood of the ear to the submaxillary, is the *parotid.* Its duct will be found passing forwards over the face to the mouth, near the angle of which it passes in through the cheek muscles.

In front of the submaxillary will be found a small gland, the *sub-lingual.*

Remove the muscles, etc., covering the larynx and trachea; cut away the front and side walls of the chest and abdomen; remove larynx, trachea, lungs, and heart.

The *gullet,* a slender muscular tube, will now be exposed in the neck; trace it through the chest; note the relative positions of the abdominal viscera as now exposed, before displacing any of them; then turning the liver up out of the way, follow the gullet in the abdomen until it ends in the stomach.

Note the form of the latter organ; its projection (*fundus*) to the left of the entry of the gullet; its *great* and *small curvatures;* its narrower *pyloric portion* on the right, from which the small intestine proceeds. Attached to the stomach, and hanging down over the other abdominal viscera, notice a thin membrane, the *omentum.*

Follow and unravel the coils of the small intestine, spreading out as far as possible the delicate membrane (*mesentery*) which slings it. In the mesentery are numerous bands of fat, running in which will be seen blood-vessels and lacteals.

The termination of the small intestine by opening into the side of the large. Observe the *cæcum* or blind end of the latter, projecting on one side of the point of entry of the small intestine; on the other side follow the large intestine until it ends at the anal aperture, cutting away the front of the pelvis to follow its terminal portion (*rectum*). The portion between the cæcum and the rectum is the *colon.*

Spread out the portion of the mesentery lying in the concavity of the first coil (*duodenum*) of the small intestine; in it will be seen a thin branched glandular mass, the *pancreas.*

Observe the *portal vein* entering the under side of the liver by several branches. Alongside it will be seen the *gall-duct,* formed by the union of two main branches, and proceeding, as a slender tube, to open into the duodenum about an inch and a half from the pyloric orifice of the stomach.

Note the *spleen:* an elongated red body lying in the mesentery, behind and to the left of the stomach.

Divide the gullet at the top of the neck, and the rectum close to the anus, and severing mesenteric bands, etc., by which intermediate portions of the alimentary canal are fixed, remove the whole tube; then cutting away the mesentery, spread it out at full length, and note the relative length and diameter of its various parts. The whole is seven or eight times as long as the head and trunk of the animal, and the small intestine forms by far the longest part of it.

Open the stomach; note that the *mucous membrane* lining the fundus is thin and smooth, and is sharply marked off from the thick corrugated mucous membrane lining the rest of the organ. (This is not the case in the human stomach.) Pass probes through the *cardiac orifice* into the gullet and through the *pyloric orifice* into the duodenum.

Remove the liver; note its general form.

Obtain from your butcher an inch or two of the small intestine of a recently killed calf. Place in 50 per cent. alcohol for twenty-four hours. Then open under water and examine with a hand lens to see the villi.

2. The Action of Saliva on Starch.—Make a thin paste of *good* arrowroot (which is almost pure starch) with boiling water. Let it cool.

a. Add two or three drops of this starch paste to half a test tubeful of cold water; next add three or four drops of solution of caustic potash and two or three drops of dilute watery solution of blue vitriol (cupric sulphate). Mix thoroughly and boil over a spirit lamp. No orange-red precipitate will result. This shows that there is no grape sugar in the starch paste.

b. Rinse the mouth thoroughly and then collect a small quantity of saliva in a test tube. Dilute with water. Add caustic potash and cupric sulphate solutions as above; mix thoroughly and boil. The mixture will become violet, but give no orange-red precipitate; therefore there is no grape sugar in saliva.

c. Take now three drops of the starch paste and a teaspoonful of saliva; mix with a half test tubeful of water. Place the mixture in a moderately warm place for five minutes. Then add a few drops of the caustic potash and cupric sulphate solutions; mix and boil. An abundant orange or brick-red precipitate will be thrown down, proving the presence of grape sugar, which has been produced by the action of the saliva on the starch.

3. Gastric Digestion.—*a.* Obtain a pig's stomach. Cut it open and wash away its contents with a gentle stream of water. Then dissect off the mucous membrane from its middle part, mince and put aside for a couple of days in three or four ounces of glycerine. The glycerine dissolves the pepsin. Then strain off the glycerine through muslin.

b. Get a butcher to " whip" some fresh drawn blood for you with a bunch of wire or twigs. The blood fibrin will collect on these (p. 181), and when thoroughly washed with water, forms a good proteid for digestion experiments. One lot of it thus obtained and washed may be put aside in 50 per cent. alcohol, and will provide material for digestion experiments for years.

c. Add a teaspoonful of muriatic acid to a pint of water.

d. Dilute a teaspoonful of the pepsin solution *a* with two tablespoonfuls of water. Fill a test tube with the mixture; add a few shreds of washed fibrin, and set aside in a warm but not *hot* place for twenty-four hours. No change will occur, showing that pepsin alone will not dissolve proteids.

e. Put some shreds of fibrin in a test tube of the mixture *c* in a warm place for twenty-four hours. The fibrin will swell up and become translucent, but will not dissolve. This shows that dilute acids will not in a short time dissolve proteids.

f. Half fill a test tube with the mixture *c*, add a teaspoonful of the pepsin solution *a*, and then a few shreds of fibrin. Place in a warm place for twenty-four hours. The fibrin will be more or less completely dissolved at the end of that time. We thus find that pepsin alone and dilute acid alone (at least in a moderate time) will not dissolve proteids, but that acting together they quickly effect a solution.

4. **The Action of Bile on Fatty Substances.**—*a.* Shake up some olive oil with water in a test tube. The two liquids soon separate when the shaking ceases.

b. Obtain an ox gall from the butcher. Cut it open and collect the bile. (The bile of herbivorous animals differs from human bile in being green in color.) Shake up some oil with bile instead of water. A creamy emulsion is formed from which the oil only slowly floats up to the top.

5. **The Action of the Pancreatic Secretion on Fats.**—*a.* Obtain a pig's pancreas; mince, and extract with about its own bulk of water for two or three hours. Strain off the watery infusion. Add to it half its bulk of oil in a test tube and shake thoroughly. The oil will be very thoroughly emulsified; and separate very slowly on standing.

6. **Action of Pancreatic Secretion on Starch.**—With some of the watery extract of pancreas perform the experiments described above under heading 2; substituting pancreatic extract for saliva.

7. **Action of Pancreatic Secretion on Proteids.**—*a.* Obtain a fresh pig's pancreas. Lay aside in a cool place for twenty-four hours. Mince, and extract for two days with twice its bulk of glycerine. Strain off the glycerine extract.

b. Dilute the glycerine extract with ten times its bulk of water. Place part of this mixture in a test tube together with some fibrin shreds, and put aside in a warm place. After twenty-four hours none of the fibrin will have been dissolved.

c. To the diluted glycerine extract as above add a teaspoonful of dilute acid (3 *c*). The fibrin will swell but not dissolve.

d. To another portion of the diluted glycerine extract add just sufficient bicarbonate of soda to make it distinctly alkaline, as tested by litmus paper. Then put in some fibrin and set aside in a warm place for a day. The fibrin will be more or less completely dissolved. We thus find that the pancreas affords a substance which, in the presence of weak alkalies, dissolves proteids.

The fat-absorbing power of the lymphatics of the small intestine is very readily demonstrable, without giving pain to an animal or any unnecessary destruction of life. In most families superfluous kittens or puppies have to be killed soon after birth. Feed a kitten or puppy on rich milk, and three hours after place it in a box or under a bell-jar with a sponge soaked with ether or chloroform. When the animal is completely insensible cut off its head, and then rapidly open the abdomen and spread out the *mesentery* (the thin membrane which slings the small intestine). In it will be seen a beautiful network of lacteal vessels filled with milk-white liquid, some of which can be collected if one of the lacteals be cut open. For comparison a kitten or puppy may be used which has had no food for eight or ten hours. The lacteals being then filled with clear, watery-looking lymph, will be recognized with difficulty.

CHAPTER XIV.

BLOOD AND LYMPH.

Why we need Blood.—Some very small animals of simple structure require no blood; every part catches its own food and gives off its own wastes to the air or water in which the creature lives. When, however, an animal is larger and more complex, made up of many organs, some of which are far away from the surface of its body, this is impossible; some organs are therefore set apart to catch food, and arrangements made to carry some of this food to the others. In our own bodies many parts lie far away from the stomach and intestines which receive, digest, and absorb our food, and from the lungs which take oxygen gas out of the air we breathe; yet every part, bone and muscle, brain and nerve, skin and gland, needs a steady supply of both of these things to keep it alive. The division of labor, in accordance with which some organs are especially set apart for the purpose of receiving substances from the outside world to build up, nourish, and repair the body, necessitates an arrangement by which the matters received shall be distributed to other parts. This distribution is accomplished by *the blood*, which flows into every organ from the crown of the head to the sole of the foot. Being pumped round

What kind of animals do not need blood? How are their wants supplied and their wastes removed? Why do we find special receptive organs in larger animals? Illustrate from the human body. What arrangement is necessitated by the fact that special organs are set apart in the body for receiving food and oxygen? How is the distribution effected?

and round, from place to place, by *the heart,* the blood picks up nourishing things in its course through the walls of the alimentary canal, and oxygen as it flows through the lungs; it then carries them to all other parts of the body.

The Removal of Wastes.—The rapidly flowing blood not only conveys a supply of nutritive material for all the organs, but is a sort of sewage stream that drains off their wastes (p. 127), and carries them to the excretory organs, by which they are sent entirely out of the body.

The blood is a middleman: on the one hand, between the receiving organs (lungs and alimentary canal) and all the rest; and on the other hand, between the excretory organs and all the others. Each part is thus kept in a well-fed and healthy state, though it may lie far distant from all places where new materials first enter the body, and from those where refuse and deleterious substances are finally passed from it.

The Blood, as every one knows, is a red liquid which is **very** widely distributed over the body, since it flows from any part of the surface when the skin is cut through. There are very few portions of the body into which blood is not carried. One of them is the outer layer of the skin;* hairs and nails, the hard parts of the teeth and most cartilages also contain no blood; these *non-vascular* tissues are

Where does the blood receive nutritive matters? Oxygen? What does it do with them?
What part does the blood play in the removal of wastes?
State briefly the functions of the blood with reference to the nutritive processes of the body
What is blood? How do we know that it is widely distributed? Name parts into which blood does not flow. How are the non-vascular tissues nourished?

---

* The absence of blood in the superficial layer of the skin may be readily shown: take a fine needle threaded with silk; by taking shallow stitches a pattern can be easily embroidered on the palm or back of the hand without drawing a drop of blood.

nourished by liquid which soaks through the walls of blood-vessels in neighboring parts.

The Histology of Blood.—Fresh blood is to the unassisted eye a red opaque liquid showing no sign of being made up of different parts; but when examined by a microscope it is

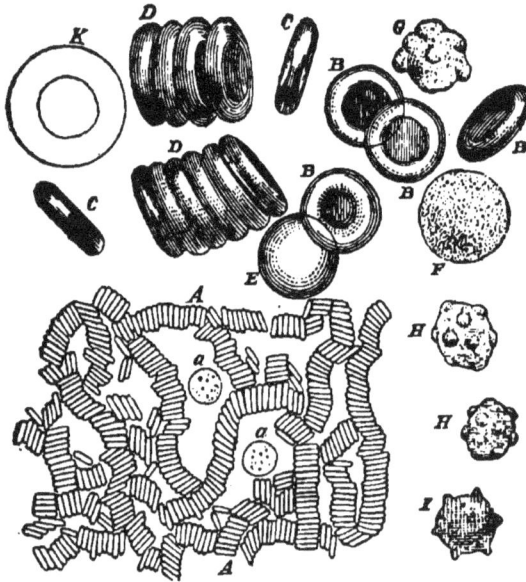

Fig. 53.—Blood-corpuscles. *A*, magnified about 400 diameters. The red corpuscles have arranged themselves in rouleaux; *a,a*, colorless corpuscles; *B*, red corpuscles more magnified and seen in focus; *E*, a red corpuscle slightly out of focus. Near the right-hand top corner is a red corpuscle seen in three-quarter face. and at *C* one seen edgewise. *F,G,H,I*, white corpuscles highly magnified.

seen to consist of a liquid, *the blood-plasma,* which has floating in it countless multitudes of closely crowded and extremely minute solid bodies known as *blood-corpuscles.* The liquid is colorless and watery-looking; the corpuscles are of two kinds, *red* and *colorless.* The red corpuscles are

Describe the appearance of fresh drawn blood. What is seen when a drop is examined with a microscope? Describe the blood-plasma Name the kinds of blood-corpuscles

by far the most numerous and give the blood its color; they are so tiny and so plentiful that about five millions of them are contained in one small drop of blood. They are so closely packed that the unaided eye cannot see the spaces between them, and so the whole blood appears uniformly red.

The **Red Corpuscles of Human Blood** (Fig. 53) are circular disks a little hollowed out on each face. Seen singly with a microscope each is not red but pale yellow; it is only when they are crowded in a heap that the mass looks red; a drop of blood spread out very thin on glass, or mixed with a tablespoonful of water, is pale yellow and not red. Soon after blood is drawn most of the red corpuscles cohere side by side in rows, something like piles of coin.

The **Red Corpuscles of other Animals.**—The red corpuscles of most mammalia resemble those of man in being circular biconcave pale yellow disks; those of camels and dromedaries, however, are oval. The blood-corpuscles of dogs are so like those of man in size that they cannot be readily distinguished; but in most cases the size is sufficiently different to enable a safe opinion to be formed. This

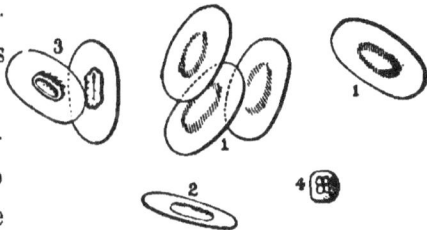

FIG. 54.—Red corpuscles of the Frog.

Which kind is most numerous? Give some idea of their number. Why does the blood look uniformly red to the unaided eye?

Describe the form of human red blood-corpuscles. What is the color of one seen by itself with a microscope? How may we show that blood looks red only when its corpuscles are crowded close together? How do the red corpuscles become arranged soon after blood is drawn?

Describe the corpuscles of most mammalia. How do those of camels and dromedaries differ from the corpuscles of other mammals? Why cannot a dog's blood be easily distinguished from human blood?

fact has often been used to further the ends of justice in determining whether spots of blood on the clothes of a suspected murderer were really due to the cause assigned by him. The red blood-corpuscles of birds, reptiles, amphibians, and fishes cannot be confounded with those of man, since they are oval and contain a nucleus in the centre such as is not found in our red corpuscles.

Hæmoglobin.—Each red corpuscle is soft and jelly like. Its chief constituent, besides water, is a substance called *hæm'-o-glo'-bin*, which has the power of combining with oxygen when in a place where that gas is plentiful, and of giving it off again in a region where oxygen is absent or present only in small quantity. Hence as the blood flows through the lungs, which are constantly supplied with fresh air, its corpuscles take up oxygen, which, as it flows on, is carried by them to distant parts of the body where oxygen is deficient, and there given up to the tissues. This oxygen-carrying is the function of the red corpuscles.

Arterial and Venous Blood.—Hæmoglobin itself is dark purplish-red in color; hæmoglobin combined with oxygen is bright scarlet red. Accordingly, the blood which flows to the lungs after giving up its oxygen is dark red in color, and that which, having got a fresh supply of oxygen, flows away from the lungs is bright scarlet. The bright red blood is called *arterial* and the dark red *venous*.

Can the blood of most mammals be certainly distinguished from human blood? Point out a use which has been made of this fact. How do the red corpuscles of birds, reptiles, and fishes differ from human?

Describe the consistence of a red corpuscle. What are its chief constituents? Point out an important property of hæmoglobin. How does this enable it to receive and distribute oxygen? What is the function of the red corpuscles?

What is the color of hæmoglobin? What of hæmoglobin combined with oxygen? What is the color of blood flowing to the lungs? Why? The color of that flowing from the lungs? Why? What is arterial blood? What venous?

The Colorless Blood Corpuscles are a little larger than the red, but much less numerous (about 1 to 300). As their name implies they contain no coloring matter. Each is a cell with a nucleus, and has the wonderful property of being able to change its own shape. Watched with a microscope the corpuscle may be seen to alter its form slowly (Fig. 55), or even to creep across the glass. These corpuscles are thus little, independently moving cells which live in our blood. The *pus* or "matter" which collects in an abscess is chiefly made up of colorless blood-corpuscles which have bored through the walls of the smallest blood-vessels. Their movements are very like those of the microscopic animal named *amœba*, and are accordingly called *amœboid*.

Fig. 55.—A white blood-corpuscle, sketched at successive intervals of a few seconds to illustrate the changes of form due to its amœboid movements.

The Coagulation of Blood.—When blood is first drawn from the living body it is perfectly liquid, flowing in any direction as readily as water. This condition is only temporary; in a few minutes the blood becomes viscid and sticky, and comes to resemble a thick red syrup; the viscidity becomes more and more marked, until, after the lapse of five or six minutes, the whole mass sets into a jelly which adheres to the vessel containing it, so that this may be inverted without any blood whatever being spilled. This stage is known as that of *gelatinization*, and is also not per-

How do the colorless corpuscles differ from the red in size and number? What is each? What property does it possess? What is seen when one is watched with the help of a microscope? What is pus? Why are the movements of the colorless corpuscles called amœboid?

What is the consistency of fresh drawn blood? What change occurs in it within a few minutes?

manent. In a few minutes the top of the jelly-like mass will be seen to be hollowed or "cupped," and in the concavity will be found a small quantity of nearly colorless liquid, the *blood-serum.* The jelly next shrinks so as to pull itself loose from the sides and bottom of the vessel containing it, and as it shrinks it squeezes out more and more serum. Ultimately we get a solid *clot,* colored red and smaller in size than the vessel in which the blood coagulated, but retaining its form, and floating in a quantity of pale yellow *serum.* The whole series of changes leading to this result is known as the *coagulation* or *clotting* of the blood.

Cause of Coagulation.—If a drop of fresh drawn blood be spread out and watched with a powerful microscope, it will be seen that its coagulation is due to the separation of very fine solid threads which run in every direction through the plasma and form a close network entangling all the corpuscles. These threads are composed of an albuminous substance known as *fibrin.* When they first form, the whole drop is much like a sponge soaked full of water (represented by the serum) and having solid bodies (the corpuscles) in its cavities. After the fibrin threads have been formed they begin to shorten; hence the fibrinous network tends to shrink in every direction, and this shrinkage is greater the longer the clotted blood is kept. At first the threads stick too firmly to the bottom and sides of the

What is meant by the stage of gelatinization? What first follows that stage? What next? What is the final result? What is the whole process called?

What is seen on watching a drop of fresh drawn blood with the aid of a good microscope? What are the separated threads composed of? To what may we compare a drop of blood in the first formation of the fibrin threads? What do the threads do after their formation?

vessel to be pulled away, and thus the first sign of the contraction of the fibrin is seen in the cupping of the surface of the gelatinized blood where the threads have no solid attachment, and there the contracting mass presses out from its meshes the first drops of serum. Finally the contraction of the fibrin overcomes its adhesion to the vessel, and the clot pulls itself loose on all sides, pressing out more and more serum. The great majority of the red corpuscles are held back in the meshes of the fibrin.

Whipped Blood.—The essential point in coagulation being the formation of fibrin in the plasma, and blood only forming a certain amount of fibrin,* if this be removed as fast as it forms the remaining blood will not clot. The fibrin may be separated by what is known as "whipping" the blood. For this purpose fresh drawn blood is stirred up vigorously with a bunch of twigs or a bundle of wire, and the sticky fibrin threads as they form adhere to these. If the twigs be then withdrawn a quantity of stringy material will be found attached to them. This is at first colored red by adhering blood-corpuscles, but by washing in water pure fibrin may be obtained perfectly white and in the form of highly elastic threads. The blood from which the fibrin has been in this way removed looks just like ordinary blood, but has lost its power of coagulating spontaneously.

Uses of Coagulation.—The living circulating blood in the healthy blood-vessels does not clot; it contains no solid

Why is the first sign of their contraction seen in the cupping? What is the final result of this contraction? Why is the clot red? How can we prevent blood from clotting? How is blood whipped? What do we find on examining the twigs after whipping blood? How may we get the pure fibrin? What are its characters? How does whipped blood differ from ordinary blood?

---

* Fibrin is formed from *fibrinogen*, a soluble albumen existing in blood-plasma.

fibrin, but this forms in it, sooner or later, when the blood
gets in any way out of the vessels or if the lining of these is
injured.   By the clotting the mouths of the small vessels
opened in a wound are clogged up, and the bleeding, which
would otherwise go on indefinitely, is stopped.   So too,
when a surgeon ties an artery, the tight ligature crushes or
tears its delicate inner surface, and the blood clots where
this is injured.   The clot becomes more and more solid,
and by the time the ligature is removed has formed a firm
plug in the cut end of the artery, which prevents bleeding.

Blood Compared with Water.—" Leaving aside its color,
we all know that blood is *thicker* than water; this is true
not only in a metaphorical but in a literal sense.   In the
first place, bulk for bulk, blood is heavier than water; ten
teaspoonfuls of blood weigh as much as ten and a half teaspoonfuls of water.   Secondly, blood contains in it solid
corpuscles and when drawn from the body forms spontaneously a solid clot, while pure water has no solid bodies
floating in it, and can only be made solid by freezing.
Thirdly, the blood liquid itself, quite apart from the corpuscles, is thicker than pure water, because it contains a
great many things dissolved in it; things which are of great
importance, because they are the foods which the blood is
carrying to, and the wastes which it is carrying from, the
various organs of the body."

The Composition of Blood-Serum.—About one half of
the bulk of fresh blood is corpuscles and the other half

When does blood clot?   Illustrate the uses of the coagulating
property of blood.
Compare blood with water, (1) as to the weight of equal bulks of the
two (specific gravity); (2) as to its microscopic structure; (3) as to its
tendency to solidify; (4) as to the composition of its plasma.   Why
are the things dissolved in the plasma of great importance?
What is the relative proportion of corpuscles and plasma in blood?

plasma. What the plasma contains we may learn by examining blood-serum, which is plasma minus fibrinogen.

Blood-serum is very different from water; if we keep on boiling pure water in a saucepan it will all go off in steam and leave nothing behind, but if we try to boil serum we find that we cannot do it; before it gets as hot as boiling water it sets into a stiff, solid mass just like the white of a hard-boiled egg. In fact the serum contains dissolved in it two albumens very like that in the white of an egg, and coagulated in a similar way by boiling. About eight and a half pounds of albuminous substances exist in one hundred pounds of blood.

Blood-serum also contains considerable quantities of oily and fatty matters, a little sugar, some common salt and carbonate of soda, and small quantities of very many other things, chiefly waste products from the various tissues. Nine tenths of the blood-plasma is water.

Composition of the Red Corpuscles.—In the fresh moist state these contain a little more than half their weight of water. Nine-tenths of their solid part is hæmoglobin ; they also contain phosphorus and iron and potassium.

The Blood Gases.—Ordinary fresh or salt water has a good deal of air dissolved in it, which fishes breathe. Blood also contains a quantity of gases which it gives off when exposed to a vacuum, about sixty pints of gas to a hundred

What may we learn by examining blood-serum? What is blood-serum?

What happens when we try to boil blood-serum? Why does it coagulate in heating? What proportion of albumen exists in blood?

What things are found in the blood-serum in addition to water? How much water is there in ten pints of blood-plasma?

How much solids do the red corpuscles contain? What proportion of these is hæmoglobin? Name other things found in the red corpuscles.

What do fishes breathe? What does blood give off when placed in a vacuum? How many pints of gas for each ten pints of blood?

pints of blood. In blood going to the lungs the main gas is carbon dioxide (or carbonic acid), which is a waste product of all the organs of the body. In blood coming from the lungs there is more oxygen.

Summary.—"Blood, then, is a very wonderful fluid: wonderful for being made up of colored corpuscles and colorless fluid, wonderful for its fibrin and power of clotting, wonderful for the many substances, for the proteids, for the ashes or minerals, for the rest of the things which are locked up in the corpuscles and in the serum.

"But you will not wonder at it when you come to see that the blood is the great circulating market of the body, in which all the things that are wanted by all parts, by the muscles, by the brain, by the skin, by the lungs, liver, and kidneys, are bought and sold. What the muscle wants it buys from the blood; what it has done with it sells back to the blood; and so with every other organ and part. As long as life lasts this buying and selling is forever going on, and this is why the blood is forever on the move, sweeping restlessly from place to place, bringing to each part the things it wants, and carrying away those with which it has done. When the blood ceases to move, the market is blocked, the buying and selling cease, and all the organs die, starved for the lack of the things which they want, choked by the abundance of things for which they have no longer any need."—*Foster.*

Hygienic Remarks.—The blood flowing from any organ will have lost or gained, or gained some things and lost

What is the most abundant gas in blood going to the lungs? What in that leaving those organs?

Why may blood be justly called a wonderful fluid? Why is its complexity not astonishing? Why is the blood always kept in movement during life? What happens when the blood ceases to move?

others, when compared with the blood which entered it. But the losses and gains in particular parts of the body are in such small proportion as, with the exception of the blood gases, to elude analysis for the most part; and, the blood from all parts being mixed up in the heart, they balance one another and produce a tolerably constant average. In health, however, the red corpuscles are present in greater proportion to the plasma after a meal than before it. Healthy sleep in proper amount also increases the proportion of red corpuscles, and want of it diminishes their number, as may be recognized in the pallid aspect of a person who has lost several night's rest. Fresh air and plenty of it favors their increase. The proportion of these corpuscles has a great importance, since they serve to carry oxygen, which is necessary for the performance of its functions, all over the body. *Anæmia* is a diseased condition characterized by pallor due to deficiency of red blood-corpuscles, and accompanied by languor and listlessness. It is not unfrequent in young girls on the verge of womanhood, and in persons overworked and confined within doors. In such cases the best remedies are open-air exercise and good food, though medicines containing iron are often of great use.

The Quantity of Blood in the Body.—The total weight of the blood is about one-thirteenth of that of the whole

What would we find on comparing the blood leaving an organ with that which entered it? What losses and gains are most easily detected? How is it that the blood maintains a tolerably uniform average composition? How does a meal affect the proportion of red corpuscles? How does sleep? Illustrate. What is the influence of plenty of fresh air? Why is the proportion of blood-corpuscles important? What is anæmia? What class of persons is apt to suffer from it? What are the best remedies for it?

What is the proportion of the weight of blood to that of the whole body?

body; a man of average size contains about twelve pounds of blood.

The Lymph.—The blood lies everywhere in closed tubes, and consequently does not come into direct contact with any of the cells which make up the body, except those which float in it and those which line the interior of the blood-vessels. At two parts of its course, however, the vessels through which it passes have extremely thin walls, and through the walls of these *capillaries* liquid transudes and bathes the various tissues. The transuded liquid is called *lymph ;* the blood makes lymph, and the lymph directly nourishes all the tis-sues except those mentioned above, with which the blood itself comes in contact.

Fig 56.—A dia-gram of a dialy-sing apparatus, containing two liquids, b and c, separated by a moist animal membrane.

Dialysis.—When two specimens of water containing different matters in solution are separated from one another by a moist animal membrane, an interchange of material will take place under certain conditions. If A be a vessel (Fig. 56), completely divided vertically by such a membrane, and a solution of common salt in water be placed on the side *b,* and a solution of sugar in water on the side *c,* it will be found after a time that some salt has got into *c* and some sugar into *b,* although there are no visible pores in the partition. Such an interchange is said to be due to *dialysis* or *osmosis,* and if the process were

How much blood is there in an average sized man?
Why does the blood not directly bathe most of the tissues? What cells come in contact with it? What are the capillaries? What is lymph? What is the nutritive function of lymph?
What happens when watery solutions of different substances are separated by a moist animal membrane? Illustrate. What is such an interchange called? What would be the result at the end of some hours?

allowed to go on for some hours the same proportions of salt and sugar would be found in the solutions on each side of the dividing membrane.

The Renewal of the Lymph.—Osmotic processes play a great part in the nutritive processes of the body. The lymph present in any organ gives up things to the cells there and gets things from them; and so, although it may have originally been tolerably like the liquid part or plasma of the blood, it soon acquires a different chemical composition. Dialysis then commences between the lymph outside and the blood inside the capillaries, and the latter gives up to the lymph new materials in place of those which it has lost, and takes from it the waste products which it has received from the tissues. When this blood, thus altered by exchanges with the lymph, reaches again the stomach and intestines, having lost some food materials, it is poorer in these than the richly supplied lymph around their cells, and takes up a supply by dialysis from it. When it reaches the excretory organs it has previously picked up a quantity of waste matters, and loses these by dialysis to the lymph there present, which is specially poor in such matters, since the excretory organs constantly deprive it of them. In consequence of the different wants and wastes of various cells, and of the same cells at different times, the lymph must vary considerably in composition in various organs of the body, and the blood flowing through them will in consequence get and lose different things in differ-

How does the lymph in an organ come to differ chemically from the blood plasma which supplied it? What results? What happens when the blood thus changed reaches stomach or intestine? What when it reaches excretory organs? Why does the lymph vary in composition in different parts of the body? How does this affect the blood?

ent places. But, receiving in its passage through one region what it loses in another, its average composition is kept pretty constant; and, through interchange with it, the average composition of the lymph also.

The Lymphatic Vessels or Absorbents.—The blood, on the whole, loses more liquid to the lymph through the capillary walls than it receives back at the same time. This de · pends mainly on the fact that the pressure on the blood inside the vessels is greater than that on the lymph outside, and so a certain amount of filtration of liquid from within out occurs through the vascular wall, in addition to the dialysis. The excess is collected from the various organs of the body into a set of *lymphatic vessels*, which carry it directly back into some of the larger blood-vessels near where these empty into the heart; and as fast as this onward flow of the lymph occurs under pressure from behind, it is renewed in the different organs, fresh liquid filtering through the capillaries to take its place as fast as the old is drained off.

Since the lymphatic vessels may be said to take up or absorb the excess of liquid drained from the blood and also the effete matters of the various organs, they are frequently called the *absorbents*.

Lacteals we have already learned to be only another name for the absorbents of the small intestine (p. 165).

How is the average composition of the blood maintained? How that of the lymph?

Give another name for the lymphatic vessels. Does the blood on the whole gain or lose liquid to the lymph as it flows through the capillaries? Explain why. What becomes of the excess of liquid drained off from the blood ?

Where do the lymphatic vessels convey it? What produces the onward flow of lymph? How is the lymph thus drained off replaced? Why are the lymphatics called absorbents?

What is meant by the lacteals?

Histology of Lymph.—Pure lymph is a colorless, watery looking liquid; examined with a microscope it is seen to contain numerous pale corpuscles exactly like those of the blood. It contains none of the red corpuscles.

Chemistry of Lymph.—Lymph is not quite so heavy as blood, though heavier than water. It may be described as blood *minus* its red corpuscles and considerably diluted, but of course in various parts of the body it will contain minute quantities of substances derived from neighboring tissues.

Summary.—To sum up: the blood and lymph provide a liquid in which the tissues of the body live; the lymph is derived from the blood, and affords the immediate nourishment of the great majority of the living cells of the body; the excess of it is finally returned to the blood, which indirectly nourishes the cells by keeping up the stock of lymph. The lymph itself moves but slowly, but it is constantly renovated by interchanges with the blood, which is kept in rapid movement by the heart, and which, besides containing a store of new food-matters for the lymph, absorbs the wastes which the various cells have poured into the latter.

What does lymph look like? What is seen when it is examined with a microscope?

How does lymph differ in density from blood and from water? How may it be briefly described? What does it contain in various regions of the body?

What do the blood and lymph provide? Whence is the lymph derived? What does it afford? What becomes of its excess? How does the blood play a part in nourishing the cells of the body? Which moves faster, lymph or blood? How is the lymph renovated? What keeps the blood in motion? What does the blood do besides renewing the food-matters in the lymph?

APPENDIX TO CHAPTER XIV.

Many of the main facts pertaining to the structure and composition of blood may be easily demonstrated as follows:

1. Kill a frog with ether (note, p. 86); cut off its head, and collect on a piece of glass a drop of the blood which flows out. Spread out the drop so that it forms a thin layer. Hold the glass up against the light, and examine the blood with a hand lens magnifying four or five diameters. The corpuscles will be readily seen floating in the plasma.

2. Wind tightly a piece of twine around the last joint of a finger, then, taking a needle, prick the skin near the root of the nail. A large drop of blood will exude. Spread it out on a piece of glass and examine, as described above for frog's blood. The corpuscles will be seen floating in the blood liquid, but not so easily as in frog's blood, since those of man are considerably smaller.

[3. If a compound microscope is available the form, size, and color of human and frog red blood-corpuscles can be demonstrated; also the tendency of the human to aggregate in rolls, and the color, form, size, and relative number of the colorless corpuscles. As any one possessing a compound microscope is sure to know how to mount a specimen of blood for examination with it, or, if not, to have at hand some treatise on the use of the microscope giving the necessary information, details need not be given here.]

4. Obtaining a large drop of human blood as above described (2) —note: a, that as it flows from the wound it is perfectly liquid; b, that it is red and very opaque; c, spread it out very thin on the glass; note that it then looks yellow when held over a sheet of white paper; d, mix a similar drop with a teaspoonful of water in a wine glass; note that the mixture is yellowish, or, if not, becomes so on further dilution.

5. Place another large drop of human blood, obtained as above indicated, on a clean glass plate. To prevent drying up cover by inverting over the drop a wine-glass whose interior has been moistened with water. In four or five minutes remove the wine-glass and note that the blood drop has set into a firm jelly. Replace the moist wine glass, and in half an hour examine again. The blood will then have separated into a tiny red clot, lying in nearly colorless serum.

6. If a slaughter-house is accessible the clotting of blood may be still better illustrated. Provide two large wide-necked glass bottles and a bundle of twigs. When the butcher bleeds an animal collect in one bottle some blood, taking care that nothing else (contents of

the stomach, for example, when the animal is bled, as is often done, by cutting off its head) gets mixed with it. Put this bottle aside until the blood clots, and carry it home with the least possible shaking. Next day the mass will exhibit a beautiful clot floating in serum. The latter will probably be tinted red, as the jolting in conveying the specimen from the slaughter house shakes some of the corpuscles out of the clot into the serum.

7. In the other bottle collect blood and beat it vigorously with the twigs for three or four minutes. Next day this specimen will not have clotted, but on the twigs will be found a quantity of stringy elastic material (fibrin), which becomes pure white when thoroughly washed with water.

8. Take some of the serum from specimen 6. Point out that it does not coagulate spontaneously. Heat it in a test tube over a spirit lamp; the albumen will be coagulated and the whole will become solid.

9. Place a small quantity of whipped blood (7) on a piece of platinum foil. Heat over a spirit lamp. After the drop dries it blackens, showing that it contains much organic matter. As the heating is continued this is burnt away, and a white ash, consisting of the mineral constituents of the blood, is left.

CHAPTER XV.

THE ANATOMY OF THE CIRCULATORY ORGANS.

The Organs of Circulation are the *heart* and the *blood-vessels;* the blood-vessels are of three kinds, *arteries, capillaries,* and *veins.* The arteries carry blood from the heart to the capillaries; the veins collect it from the capillaries and return it to the heart. There are two distinct sets of blood-vessels in the body, both connected with the heart; one set carries blood to, through, and from the lungs, the other guides its flow through all the remaining organs; the former are known as the *pulmonary,* the latter as the *systemic* blood-vessels.

General Statement.—During life the pumping of the heart keeps the blood flowing rapidly through the paths marked out for it by the blood-vessels ; these paths it never leaves except in cases of disease or injury.

The blood-vessels form a continuous system of closed tubes comparable in a certain way to the water-mains of a city. These tubes begin at the heart, and are very much branched except close to it, just as the water pipes are single only where the main aqueduct leaves the reservoir,

Name the organs of circulation. Name the kinds of blood-vessels. What is the function of the arteries? Of the veins? How many sets of blood-vessels are there in the body? What does each set do? What are they called?

What is brought about by the beat of the heart? Under what circumstances may blood leave a blood channel ?

What do the blood-vessels form ? Illustrate their function by comparison with the water-mains of a city.

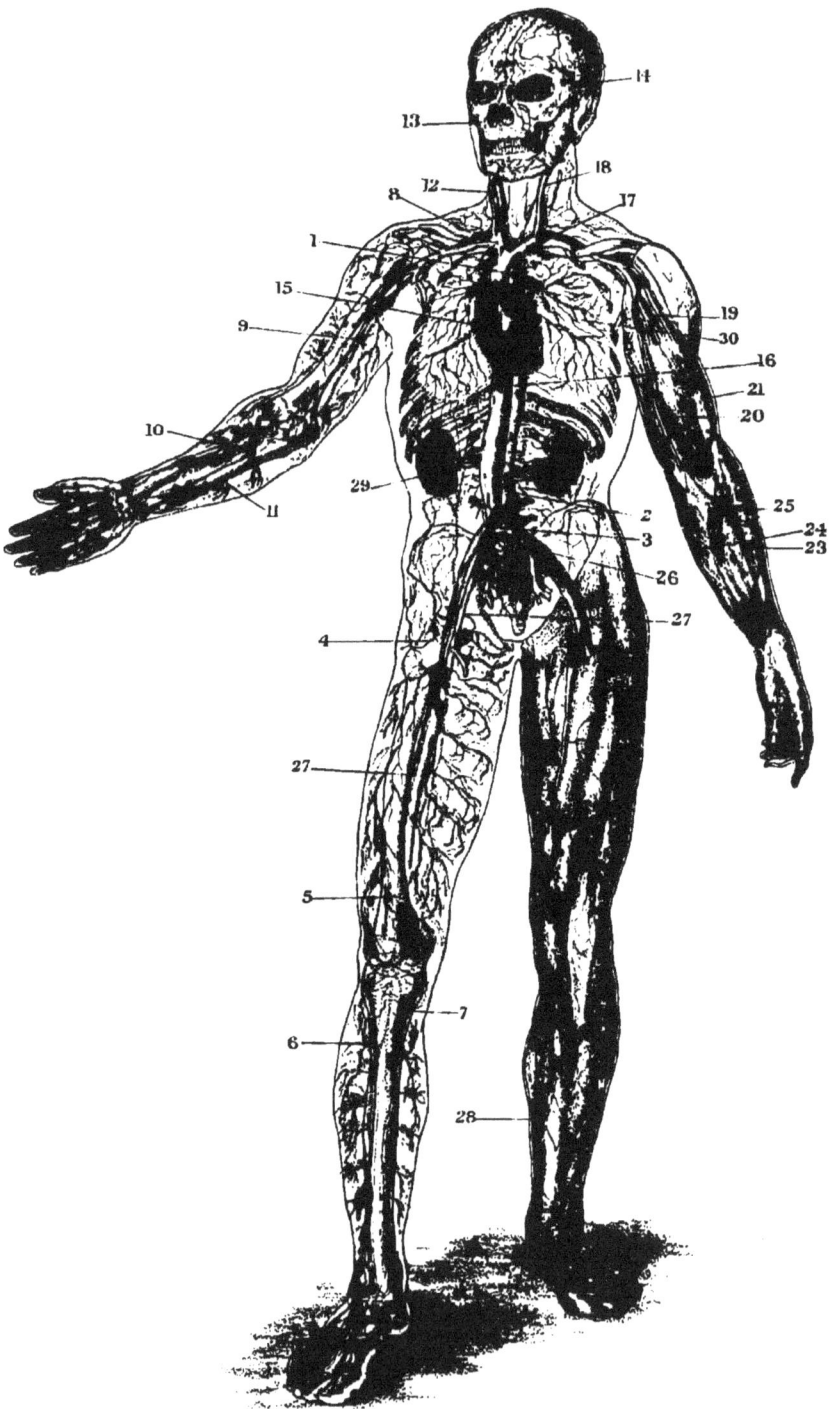

PLATE IV.—THE CHIEF ARTERIES AND VEINS OF THE BODY.

EXPLANATION OF PLATE IV.

THE CIRCULATORY ORGANS.

The arteries (except the pulmonary) and the left side of the heart are colored red; the veins, (except the pulmonary) and the right half of the heart blue: on the limbs of the left side the arteries are omitted and only the superficial veins are shown.

1. Aorta, near its origin from the left ventricle of the heart.
2. Lower end of aorta.
3. Iliac artery.
4. Femoral artery.
5. Popliteal artery; the continuation of the femoral which passes behind the knee-joint.
6, 7. The main trunks (anterior and posterior tibial arteries into which the popliteal divides).
8. Subclavian artery.
9. Brachial artery.
10. Radial artery.
11. Ulnar artery.
12. Common carotid artery.
13. Facial artery.
14. Temporal artery.
15. Right side of Heart, with superior vena cava joining it above, and inferior vena cava (16) passing up to it from below.
17. Innominate vein, formed by the union of subclavian and jugular veins. The right and left innominate veins unite to form the superior cava.
18. Left internal jugular vein.
19. Axillary vein.
20. Basilic vein.
22. Cephalic vein.
23. Radial vein.
24. Ulnar vein.
25. Median vein.
26. Iliac vein.
27. Femoral vein.
28. Long saphenous vein.
29. The kidney; attached to it are seen the renal artery and vein.
30. Branches of the pulmonary arteries and veins in the lung.

and more and more divided the further one follows them from that point, until, in the various houses, they end in numerous but very much smaller tubes. The course of the blood differs, however, essentially from that of the water supplied to a city, for the water does not return to the reservoir, whereas the blood is carried back to the heart: instead of having a large supply of liquid stored up as in a reservoir, there is at any one time only quite a small amount in the heart, but this is steadily replaced by the inflow through the veins as fast as it is carried off by outflow through the arteries.

General Functions of the Different Parts of the Vascular System.—The blood vascular system is quite closed except at two points, one on each side of the neck, where lymph vessels pour the excess of lymph back into the veins. Valves at these two points let lymph flow into the blood-vessels, but will not allow blood to pass the other way. Accordingly everything which leaves the blood must do so by oozing through the walls of the blood-vessels, and everything which enters it must do the same, except matters conveyed in by the lymph at the points above mentioned. This interchange through the walls of the vessels takes place only in the capillaries. In India rain falls only at certain seasons, and is stored in huge tanks ; during subsequent dry months the water is distributed over the fields by a set of small ditches and channels, cut through them in all directions, and from which the liquid soaks through the sur-

How does the blood flow differ from that of water in the mains of a city? How is the supply of blood in the heart kept up?

Where is the blood vascular system not closed? What occurs at its openings? What is the function of the valves at these openings? How must substances leave the blood? How enter it? In what vessels does interchange through the walls of the blood-vessels occur? Illustrate the function of the capillaries by comparison with an irrigation system.

rounding soil; this is known as irrigation, and the capillaries may be said to form an *irrigation system* for the body. In a certain sense also they may be compared to the water spigots of a house, which lie at the end of the supply tubes, the arteries; but, to make the comparison more accurate, we would have to imagine instead of ordinary spigots bags of very fine muslin through which the water oozed when the tap was turned on. The capillaries, though far the smallest tubes in the vascular system, are the really important parts; the heart, arteries, and veins are all merely arrangements for keeping the capillaries full, and renewing the blood within them. It is while flowing through these and soaking through their walls that the blood does its physiological work.

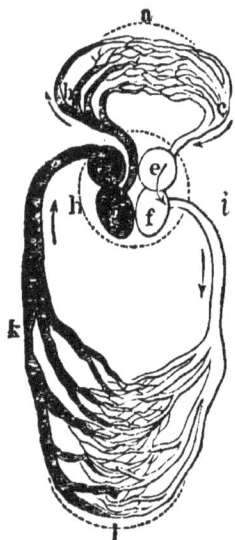

Diagram of the Circulatory Organs.— The general relationship of heart, arteries, capillaries, and veins may be gathered from the accompanying diagram (Fig. 57). The heart is essentially a bag with muscular walls. and internally divided into four chambers (*d, g, e, f*). Those at one end (*d* and *e*) receive blood from vessels opening into them and known as the *veins*. From there the blood passes on to the remaining chambers (*g* and *f*), which have very muscular walls and, forcibly contracting, drive the blood out into

Fig. 57.—The heart and blood-vessels diagrammatically represented.

By comparison with the water pipes of a house. What is the use of heart, arteries, and veins? When does the blood do its physiological work?

Sketch on the blackboard a diagram of the circulatory organs. What is briefly the structure of the heart? Where does the blood enter the heart? From what vessels? Where does it go next?

vessels (*i* and *b*) which communicate with them and are known as the *arteries*. The big arteries divide into smaller; these into smaller again (Plate IV), until the branches become too small to be traced by the unaided eye, and these smallest branches end in the *capillaries*, through which the blood flows and enters the commencements of the veins; the veins convey it again to the heart. At certain points in the course of the blood-path valves are placed, which prevent a back-flow. This alternating reception of blood at one end of the heart and its ejection from the other occurs about seventy times a minute during health.

The Position of the Heart.—The heart (*h*, Fig. 4) lies in the chest, immediately above the diaphragm and opposite the lower two thirds of the breast-bone. It is conical in form, with its *base* or broader end turned upwards and projecting a little on the right of the sternum, while its narrow end or *apex*, turned downwards, projects to the left of that bone, where it may be felt beating between the cartilages of the fifth and sixth ribs. The position of the organ in the body is, therefore, oblique. It does not, however, lie on the left side, as is so commonly believed, but very nearly in the middle line, with the upper part inclined to the right, and the lower (which may be more easily felt beating—hence the common belief) to the left.

The Pericardium.—The heart does not lie bare in the

What then happens to it? How are the small arteries formed? In what vessels do they end? What becomes of blood which has flowed through the capillaries? Where do the veins carry it? How is back-flow prevented? How frequently does the heart receive and pump out blood?

Where is the heart situated? What is its form? Where does its base lie? Where its apex? Where may we feel the apex beat? What is the origin of the common belief that the heart is on the left side?

chest, but is surrounded by a loose bag composed of connective tissue and called the *pericardium*. This bag, like the heart, is conical but turned the other way, its broad part being lowest and attached to the upper surface of the diaphragm. Internally it is lined by a smooth *serous membrane* like that lining the abdominal cavity, and a similar layer (the *visceral layer* of the pericardium) covers the outside of the heart itself, adhering closely to it. In the space between the serous membranes is a small quantity of liquid which moistens the contiguous surfaces, and diminishes the friction which would otherwise occur during the movements of the heart.

Suppose a pear put in a bag of about the same shape, but larger, and turned the other way so that the big end of the bag was round the small end of the pear; then you will get a good idea of how the pericardium lies with reference to the heart. If the outside of the pear and the inside of the bag were covered with paint, this would represent the serous membrane, and a few drops of water between the pear and the bag would represent the serous liquid. To complete the comparison we may imagine the pear to have eight or nine stalks which reached out from it through the bag; these would answer to the blood-vessels entering and leaving the heart.

Note.—Sometimes the pericardium becomes inflamed, this affection being known as *pericarditis*. It is extremely apt to occur in rheumatic fever, and extreme care should be taken never, even for a moment, except under medical

What is the pericardium? What is its form? What lines it? What covers the outside of the heart? What lies between the visceral layer of the pericardium and the outer bag? What is its use?
Illustrate the relations of heart and pericardium.
What is pericarditis? In what disease is it especially apt to occur?

advice, to expose a patient to cold during that disease, since any chill is then especially apt to set up pericarditis. In the earlier stages of pericardiac inflammation the rubbing surfaces on the outside of the heart and the inside of the pericardium become roughened, and their friction produces a sound which can be heard with a stethoscope. In later stages great quantities of liquid may accumulate in the pericardium so as to seriously impede the heart's beat.

The Cavities of the Heart.—On opening the heart (see diagram, Fig. 57, p. 222) it is found to be subdivided by a longitudinal partition or *septum* into completely separated right and left halves, the partition running from about the middle of the base to a point a little on the right of the apex. Each of the chambers on the sides of the septum is again incompletely divided transversely into a thinner basal portion into which veins open, known as the *auricle,* and a thicker apical portion from which arteries arise and called the *ventricle.* The heart cavity thus consists of a right auricle and ventricle and a left auricle and ventricle, each auricle communicating by an *auriculo-ventricular orifice* with the ventricle on its own side; there is no direct communication whatever through the septum between the opposite sides of the heart. To get from one side to the other the blood must leave the heart and pass through a set of

Why should a person with rheumatic fever never be exposed to cold except under skilled advice? What can be heard with a stethoscope in the early stages of pericarditis? What may occur in its later stages?

What is seen on opening the heart? In what direction does the septum run? What is an auricle? A ventricle? Enumerate the chambers of the heart cavity. With what does each auricle communicate? Is there a direct connection between the right and left sides of the heart? What must the blood do to get from one side of the heart to the other?

capillaries, as may readily be seen by tracing the course of the vessels in Fig. 57.

The Vessels connected with the Different Chambers of the Heart.—One big artery, called the *aorta*, springs from the left ventricle; it runs back to the pelvis after giving off very many branches on its way and then divides into an artery for each leg. Its big branches divide into smaller and these into smaller again, until they become too small to be traced by the unaided eye. They spread through the whole body, to muscles, and bones, and skin, and brain, and stomach, and intestines, and liver, and kidneys; they finally end in the systemic capillaries. The systemic veins collect the blood from the capillaries of the different organs, and all these veins

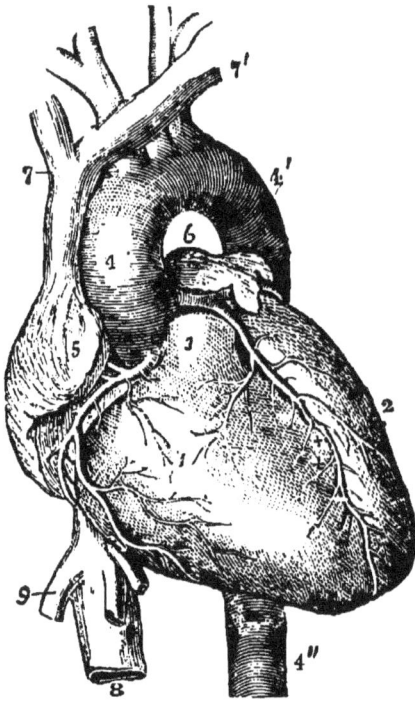

Fig. 58.—View of the Heart and Great Vessels from before.

The pulmonary artery has been cut short close to its origin. 1, right ventricle; 2, left ventricle; 3, root of the pulmonary artery; 4, 4′, arch of the aorta; 4″, the descending thoracic aorta; 5, part of the right auricle; 6, part of the left auricle; 7, 7′, innominate veins joining to form the vena cava superior; 8, inferior vena cava; 9, one of the large hepatic veins; ×, placed in the right auriculo-ventricular groove, points to the right or posterior coronary artery; ×, ×, placed in the anterior interventricular groove, indicate the left or anterior coronary artery.

What artery arises from the left ventricle? To what point does it run? What does it give off on its course? How does it end? What becomes of its branches? In what do they end? What vessels collect the blood from the systemic capillaries?

unite to form the superior **and inferior** *venæ cavæ* or
the upper and lower hollow veins. These carry the blood
to the right auricle; thence it enters the right ventricle
from which arises one vessel, *the pulmonary artery;* this
divides into a branch for each lung; each branch splits up
into minute arteries in its own lung, and these end in the
pulmonary capillaries. From the pulmonary capillaries the
blood of each lung is collected into two *pulmonary veins,*
and the four pulmonary veins open into the left auricle.

Summary.—One artery, the aorta, arises from the left
ventricle. The blood carried out by the aorta comes back
by the upper and lower venæ cavæ to the right auricle; this
blood then goes to the right ventricle and is sent thence
through the pulmonary artery, which splits up into branches
for the lungs. The blood, carried out by the pulmonary
artery from the right ventricle of the heart, returns to the
left auricle by four pulmonary veins, two from each lung;
and then enters the left ventricle and begins its flow again
through the aorta.

How the Heart is Nourished.—The heart is a very hard-
worked organ, and needs an abundant supply of nourishment. Its walls are much too thick to allow this to soak
in sufficient abundance all through them, from the blood
flowing through its cavities; accordingly they are perme-
ated by a very close network of capillary blood vessels.
These are supplied by the *right* and left *coronary arteries*

Where do the venæ cavæ carry it? Where does it pass from the
right auricle? What vessel arises from the right ventricle? Into
what does the pulmonary artery divide? What happens to each
branch? How do the branches finally end? Into what is the blood
which flows through the pulmonary capillaries collected? How
many pulmonary veins are there? Where do they end?
State briefly the course of the blood flow.

(Fig. 58), which are the two very first branches of the
aorta, and the blood from them is collected by the *coronary
veins* and poured by them directly into the right auricle.

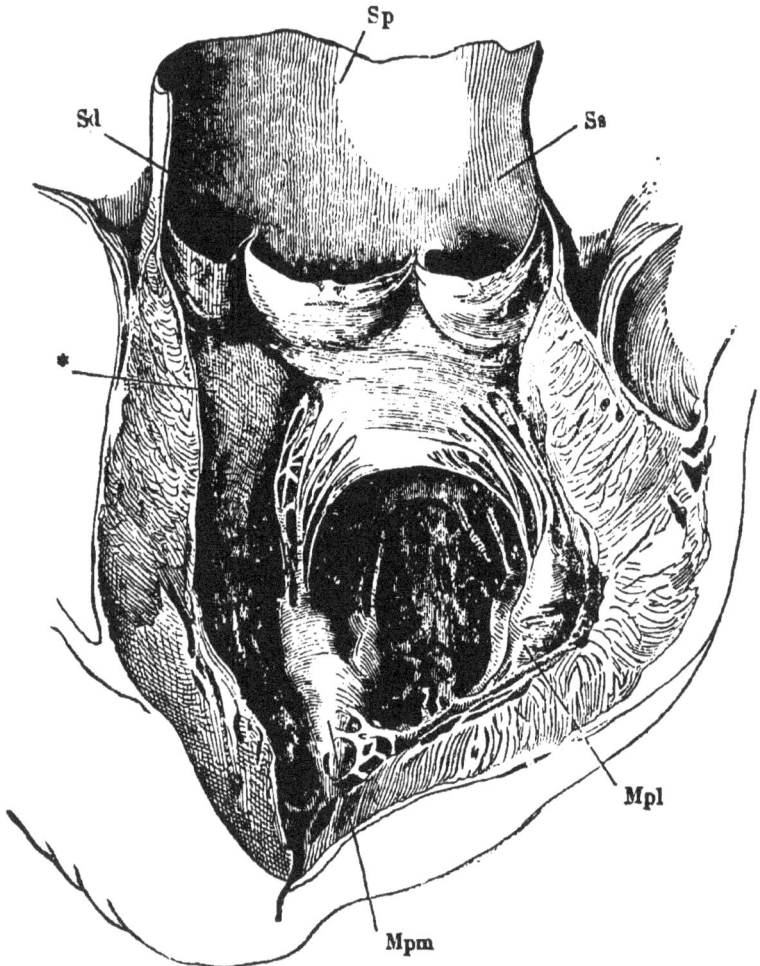

FIG. 59.—The left ventricle and the commencement of the aorta laid open. *Mpm,
Mpl.* the papillary muscles. From their upper ends are seen the chordæ tendineæ
proceeding to the edges of the flaps of the mitral valve. The opening into the
auricle lies between these flaps. At the beginning of the aorta are seen its three
pouch-like semilunar valves.

What are the coronary arteries ?   The coronary veins ?

The Auriculo-Ventricular Valves.—Between each auricle of the heart and the ventricle of the same side are found valves which allow blood to pass from the auricle to the ventricle, but prevent any flow in the opposite direction. These valves are known as the *tricuspid* and *mitral* valves. The mitral valve (Fig. 59) consists of two flaps fixed by their bases to the margins of the opening between the left auricle and the left ventricle; their edges hang down into the ventricle when the heart is empty. These edges are not free, but have fixed to them a number of stout connective-tissue cords, the *chordæ tendineæ*, which are fixed below to muscular elevations, the *papillary muscles, Mpm* and *Mpl*, on the interior of the ventricle. The cords are long enough to let the valve flaps rise into a horizontal position and so to close the opening between auricle and ventricle, which lies behind the opened aorta, *Sp*, represented in the figure. The *tricuspid valve* is like the mitral, but with three flaps instead of two.

Semilunar Valves.—These are six in number; three at the mouth of the aorta, Fig. 59, and three, quite like them, at the mouth of the pulmonary artery. Each is a strong crescentic pouch fixed by its more curved border, and with its free edge turned away from the heart. When the valves are in action their free edges meet across the vessel and prevent blood from flowing back into the ventricle.

The Course of the Main Arteries of the Body (Fig. 60). —The aorta after leaving the left ventricle makes an arch

What is found between each auricle and ventricle? What are they called? Describe the mitral valve. Where is it placed? What are the chordæ tendineæ? The papillary muscles? How far will the cords allow the valve flaps to rise? How does the tricuspid valve differ from the mitral?

How many semilunar valves are there? Where are they placed? Describe them. What is their use?

(*a*A) with its convexity towards the head. From the heart end of this arch arise the *coronary arteries,* which carry blood into the walls of the heart. From the convexity of the arch spring the three large trunks: the *innominate artery ;* the *left common carotid artery* (*c*s); and the *left subclavian artery* (s*si*). The innominate soon divides into the *right subclavian* (s*d*), and the *right common carotid* (c*d*) Each common carotid runs up the neck on its own side, and divides into branches for the neck, face, scalp, and brain. Each subclavian continues across the arm-pit as the *axillary artery* (Λ*x*), and then runs down to near the elbow as the *brachial artery* (B). Just above the elbow it divides into the *radial* and *ulnar arteries* (R, U,) which supply the fore-arm, and end in small branches for the hand

Beyond its arch the aorta runs back close to the spinal column as the *thoracic aorta* (Λ*t*), which gives off branches to the walls of the chest and some organs in that cavity. The vessel then passes through the diaphragm, and continues as the *abdominal aorta* (Λ*ab*) to the lower part of the abdomen. The main branches of the abdominal aorta are: (1) the *cœliac axis,* which divides into branches for the stomach, liver, and spleen; (2) the *upper* and *lower mesenteric arteries,* which supply the intestines with blood; (3) the *renal arteries* (K), which carry blood to the kidneys.

What is meant by the aortic arch? What are its first branches? What branches are given off from the upper side of the aortic arch? Into what vessels does the innominate artery divide? To what parts do the common carotid arteries carry blood? In what does the subclavian artery terminate? What vessels supply the fore-arm with blood? How do they end?

What is the thoracic aorta?

The abdominal aorta? Name the chief branches of the abdominal aorta. What organs are supplied by the cœliac axis? The mesenteric arteries? The renal arteries?

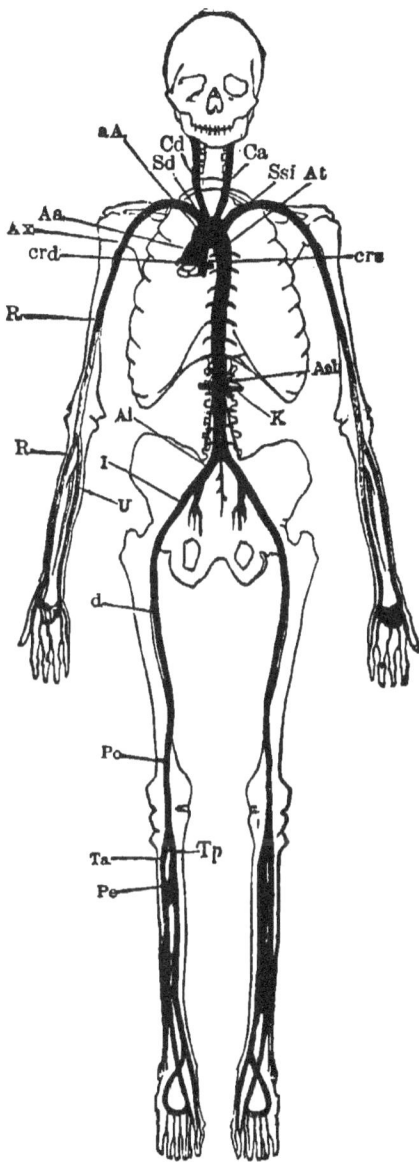

FIG. 60.—The main arteries of the body. *Crd* and *Crs*, right and left coronary arteries of the heart, cut short near their origin; *Aa*, and *aA*, aortic arch; *At*, thoracic aorta; *Aab*, abdominal aorta; *K*, renal artery; *Sd*, right, and *Ssi*, left subclavian; *Cd*, right, and *Cs*, left carotid; *Ax*, axillary artery; *B*, brachial artery; *U*, ulnar artery; *R*, radial artery; *Ai*, common iliac artery; *I*, external iliac artery; *C*, femoral artery; *Po*, popliteal artery; *Ta*, anterior, and *Tp*, posterior tibial artery; *Pe*, peroneal artery.

The two trunks into which the posterior end of the abdominal aorta divides (*Ai*) are named the *common iliac arteries;* each gives off some branches in the pelvis, and then continues along the thigh as the *femoral artery* (*C*); this runs to the knee-joint, behind which it is called the *popliteal artery* (*Po*). The popliteal artery divides into the *peroneal* (*Pe*) and *tibial* (*Ta, Tp*) arteries, which supply the lower leg and the foot.

The Properties of the Arteries.—Two fundamental facts must be borne in mind in connection with the arteries: *First,* that they are highly elastic and extensible; a large artery is, in this respect, much like a piece of rubber tubing of the same size. *Second,* the arteries have rings of muscular tissue in their walls, and when the muscle contracts the bore of the artery (and consequently the amount of blood which flows through it) is diminished. When the muscle relaxes, the bore of the artery is increased, and more blood passes along it to the capillaries in which it ends.

The Capillaries.—The smallest arteries pass into the *capillaries,* which have very thin walls, and form very close networks in nearly all parts of the body; their immense number compensating for their smaller size. The average diameter of a capillary vessel is so small that only two or three blood-corpuscles can pass through it abreast, and in many parts the capillaries lie so close together that a pin's

Into what vessels does the abdominal aorta ultimately divide? What is the femoral artery derived from? To what point does it run? What is the popliteal artery? Into what branches does the popliteal artery divide? What parts do they supply with blood?

What main facts are to be borne in mind in connection with the arteries? How is the quantity of blood which an artery will let pass regulated?

What is found when the arteries are followed to the ends of their smallest branches? Describe the structure, arrangement and size of the capillaries.

point could not be inserted between two of them, as, for instance, in the deep layers of the skin which can hardly be pricked anywhere with a needle without drawing blood. It is while flowing in these delicate tubes that the blood does its nutritive work, the arteries being merely supply-tubes for the capillaries, through whose delicate walls liquid containing nourishment exudes from the blood to bathe the various tissues. Imagine a piece of the finest lace, with all its threads consisting of hollow tubes, and diminished twenty times in size, and you will have some idea of the capillaries.

The Veins.—The first veins arise from the capillary net-works in various organs, and like the last arteries are very small. They soon increase in size by union, and so form larger and larger trunks alongside the main artery of the part, but there are many more large veins just beneath the skin than there are large arteries. This is especially the case in the limbs, the main veins of which are superficial, and can in many persons be seen as faint blue lines through the skin.

Why the large Arteries usually lie deep.—The heart pumps the blood with great force into the arteries, and if an artery is cut very rapid and dangerous bleeding occurs ; the veins, if cut, do not bleed nearly so violently as an artery of the same size. Hence it is less dangerous to have a large vein than a large artery close under the skin.

Point out a fact illustrating the closeness of the capillaries in many parts of the body. What does the blood do as it flows through the capillaries?

Where do the first veins arise? What is their size? How do they increase in size? Where do the larger veins lie? In what parts of the body do we especially find large veins close beneath the skin?

Why are the large arteries, as a rule, placed deeper than veins of corresponding size?

The Valves of the Veins.—Except the pulmonary artery and the aorta, which have the semilunar valves, arteries have no valves. Most veins, on the contrary, contain many valves formed by pouches of their lining, which resemble in form the semilunar valves of the aorta and the pulmonary artery,

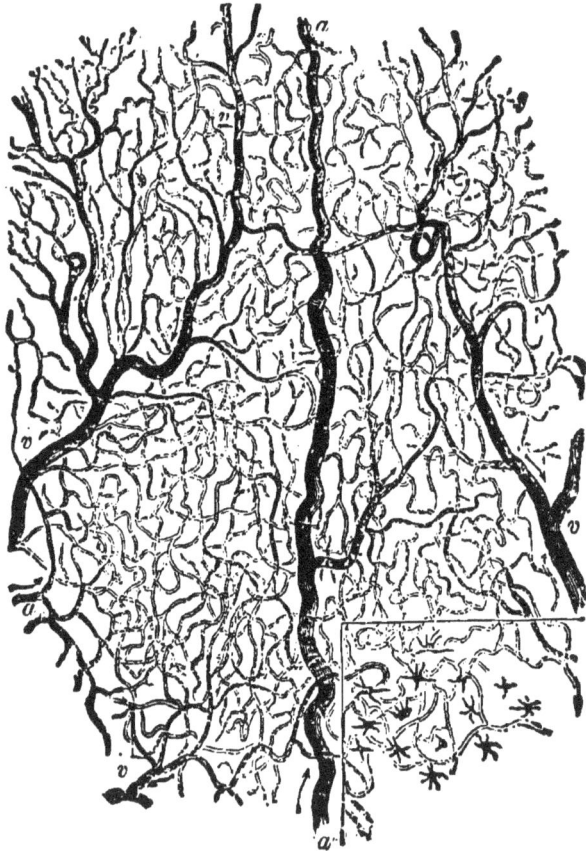

FIG. 61.—A small portion of the capillary network as seen in the frog's web when magnified about 25 diameters. *a*, a small artery feeding the capillaries; *v*, *v*, small veins carrying blood back from the latter.

What arteries have valves? Where are these valves placed? How do veins differ from arteries in regard to the presence of valves in them?

but are turned in the opposite direction, having the edge nearest the heart free and the other fixed. These valves permit blood to flow only towards the heart, for a current in that direction, as in the upper diagram, Fig. 62, presses the valve close against the side of the vessel, and meets with no obstruction from it. Should any back-flow be attempted, however, the current closes up the valve and bars its own passage, as indicated in the lower figure. These valves are most numerous in superficial veins and those of muscular parts. Usually the vein is a little dilated opposite a valve, and hence in parts where the valves are numerous gets a knotted look. On tying a cord tightly round the fore-arm, so as to stop the flow in its subcutaneous veins and cause their dilatation, the points at which valves are placed can be recognized by their swollen appearance. The valves are most frequently found where two veins communicate.

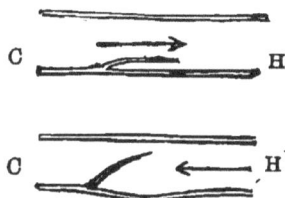

Fig. 62.—Diagram to illustrate the mode of action of the valves of the veins. C. the capillary, *H*, the heart end of the vessel.

The Course of the Blood.—From what has been said it is clear that the movement of the blood is a *circulation*. Starting from any one chamber of the heart it will in time return to it; but to do this it must pass through at least two sets of capillaries; one of these is connected with the

In what direction do the valves of the veins allow the blood to pass? Make a diagram illustrating the action of the valves. In what veins are the valves most numerous? Why does a vein with many valves appear knotted? How may we see the dilatations of the veins opposite the valves? Where are the valves of the veins most frequently placed?

If we followed the blood course steadily from one chamber of the heart what would we find in time? Through what must blood pass before returning to the chamber of the heart which it leaves?

aorta, and the other with the pulmonary artery, and in its circuit the blood returns to the heart twice. Leaving the left side it returns to the right, and leaving the right it returns to the left; and there is no road for it from one side of the heart to the other except through a capillary network. Moreover, it always leaves *from* a ventricle through an artery, and returns *to* an auricle through a vein.

There is then really only one circulation; but it is not uncommon to speak of two, the flow from the left side of the heart to the right through most of the body being called the *systemic* or *greater circulation,* and from the right to the left through the lungs the *pulmonary* or *lesser circulation.* But since, after completing either of these alone, the blood is not again at the point from which it started, but is separated from it by the septum of the heart, neither is a "circulation" in the proper sense of the word, for a circulation implies that any object at the end of its course is again exactly where it was at the commencement.

The Portal Circulation.—A certain portion of the blood which leaves the left ventricle of the heart through the aorta has to pass through three sets of capillaries before it can again return there. This is the portion which goes through the stomach and intestines. After traversing the capillaries of those organs it is collected into the *portal*

Through what does blood always leave the heart? To what does it return? How many *circulations* are there really? What is meant by the systemic circulation? What by the pulmonary? Why is neither a true "circulation" in the proper sense of the word?

How many sets of capillaries does some blood pass through in a complete circulation? What portion of the blood is it? What vessel does it enter after traversing the capillaries of the stomach and intestines?

\

vein, which enters the liver, and breaking up there into finer and finer branches like an artery, ends in the capillaries of that organ, forming the second set which this blood passes through on its course. From these it is collected by the *hepatic veins,* which pour it into the inferior vena cava which carries it to the right auricle, so that it has still to pass through the pulmonary capillaries to get back to the left side of the heart. The portal vein is the only one in the human body which thus like an artery feeds a capillary network, and the flow from the stomach and intestines through the liver to the inferior vena cava, is often spoken of as the *portal circulation.*

Diagram of the Circulation.— Since the two halves of the heart, although placed in proximity in the body, are actually completely separated from one another by an impervious partition, we may con-

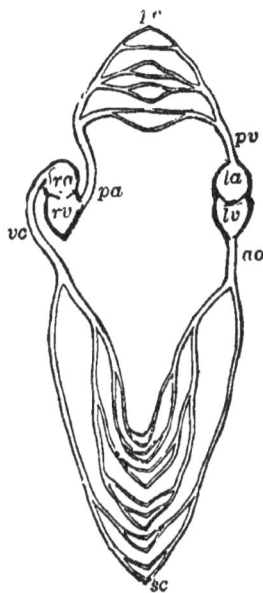

FIG. 63 Diagram of the blood vascular system, showing that it forms a single closed circuit with two pumps in it, represented by the right and left halves of the heart, which are separated in the diagram. *ra* and *rv,* right auricle and ventricle; *la* and *lv,* left auricle and ventricle; *ao,* aorta; *sc,* systemic capillaries; *vc,* venæ cavæ; *pa,* pulmonary artery; *pc,* pulmonary capillaries; *pv,* pulmonary veins.

veniently represent the course of the blood as in the accompanying diagram (Fig. 63), in which the right and

Where does this vein carry it? In what does it end? Into what vessels is the blood of the capillaries of the liver collected? Where do they convey it? What chamber of the heart does it first reach? Through what must it pass to get back to the left ventricle? How does the portal vein differ from all others in the body? What is meant by the portal circulation?

left halves of the heart are represented at different points in the vascular system. Such a diagram makes it clear that the heart is really two pumps working side by side, and each engaged in forcing blood to the other. Starting from the left auricle, *la,* and following the flow, we trace it through the left ventricle, and along the branches of the aorta into the systemic capillaries, *sc;* thence it passes back through the systemic veins, *vc.* Reaching the right auricle, *ra,* it is sent into the right ventricle, *rv,* and thence through the pulmonary artery, *pa,* to the lung capillaries, *pc,* from which the pulmonary veins, *pv,* carry it to the left auricle, which drives it into the left ventricle, *lv,* and this again into the aorta.

Arterial and Venous Blood.—The blood when flowing in the pulmonary capillaries gives up carbon dioxide (a waste product which it has gathered in its flow through the other organs) to the air, and receives oxygen from it; since its coloring matter (hæmoglobin) forms a scarlet compound with oxygen, the blood which flows to the left auricle through the pulmonary veins is of a bright red color. This color it maintains until it reaches the systemic capillaries, but in these it loses much oxygen to the surrounding tissues, and gains much carbon dioxide from them. But the blood-coloring matter which has lost its oxygen has a dark purple-black color, and since this unoxidized or "reduced" hæmoglobin is now in excess, the

Why are we justified in diagrammatically representing the heart as made of two separated parts? Starting from the left auricle, describe the course of the blood until it returns there.
What does the blood give up in the pulmonary capillaries? What does it receive? Why is it bright red when it enters the left auricle? How far in its course does it keep this color? What gases does the blood gain and lose in the systemic capillaries?

blood returns to the right auricle of the heart by the venæ cavæ of a dark purple-red color. This color it keeps until it reaches the lungs, where the reduced hæmoglobin becomes again oxidized. The bright red blood, rich in oxygen and poor in carbon dioxide, is known as "arterial blood," and the dark red as "venous blood;" and it must be borne in mind that the terms have this peculiar technical meaning, and that the pulmonary *veins* contain *arterial* blood, and the pulmonary *arteries* contain *venous* blood. The change from arterial to venous takes place in the systemic capillaries, and from venous to arterial in the pulmonary capillaries.

What color is the blood when returned to the right auricle? Why? What is meant by arterial blood? By venous? What veins contain arterial blood? What arteries venous? Where does the change from arterial to venous occur? Where that from venous to arterial?

APPENDIX TO CHAPTER XV.

1. In the following directions "dorsal" means the side of the heart naturally turned towards the vertebral column, "ventral" the side next the breast-bone; "right" and "left" refer to the proper right and left of the heart when in its natural position in the body; "anterior" means more towards the head in the natural position of the parts; and "posterior" the part turned away from the head.

2. Get your butcher to obtain for you a sheep's heart, not cut out of the bag (pericardium), and still connected with the lungs. Impress upon him that no hole must be punctured in the heart, such as is usually made when a slaughtered sheep is cut up for market.

3. Place the heart and lungs on their dorsal sides on a table in their normal relative positions, and with the windpipe directed away from you. Note the loose bag (*pericardium*) in which the heart lies, and the piece of midriff (*diaphragm*) which usually is found attached to its posterior end.

4. Carefully dissecting away adherent fat, etc., trace the vessels

below named until they enter the pericardium. Be very careful not to cut the veins, which, being thin, collapse when empty, and may be easily overlooked until injured. As each vein is found stuff it with raw cotton, which makes its dissection much easier.

a. The *vena cava inferior :* find it on the under (abdominal) side of the diaphragm; thence follow it until it enters the pericardium, about three inches further up; to follow it in this part of its course, turn the right lung towards your left and the heart towards your right.

The vein just below the diaphragm may be seen to receive several large vessels, the *hepatic veins.*

As it passes through the midriff, two veins from that organ enter it.

Between diaphragm and pericardium the inferior cava receives no branch; but, lying on its left side, will be seen the lower end of the *right phrenic nerve,* ending below in several branches to the diaphragm.

b. Superior vena cava : seek its lower end, entering the pericardium about one inch above the entry of the inferior cava; thence trace it up to the point where it has been cut across; stuff and clean it.

c. Between the ends of the two venæ cavæ will be seen the two *right pulmonary veins,* proceeding from the lung and entering the pericardium; clean and stuff them.

5. Turn the right lung and the heart back into their natural positions; clear away the loose fat in front of the pericardium, and seek and clean the following vessels in the mass of tissue lying anterior to the heart, and on the ventral side of the windpipe.

a. The *aorta :* immediately on leaving the pericardium this vessel gives off a large branch; it then arches back and runs down behind the heart and lungs, giving off several branches on its way.

b. The *pulmonary artery :* this will be found imbedded in fat on the dorsal side of the aorta. After a course, outside the pericardium, of about an inch, it ends by dividing into two large branches (right and left pulmonary arteries), which subdivide into smaller vessels as they enter the lungs.

c. Observe the thickness and firmness of the arterial walls as compared with those of the veins; they stand out without being stuffed.

6. Notice, on the ventral side of the left pulmonary artery, the *left pulmonary veins* passing from the lung into the pericardium.

7. Up to this point the dissection may be made before the meeting of the class; on the preparation demonstrate the anatomical facts above noted and then proceed as follows:

8. Slit open the pericardiac bag, and note its smooth, moist, glistening inner surface, and the similar character of the outer surface of the heart. Cut away the pericardium carefully from the entrances of the various vessels which you have already traced to it. As this

is done, you will notice that inside the pericardium the pulmonary artery lies on the ventral side of the aorta.

9. Note the general form of the heart—that of a cone with its apex turned towards the diaphragm. Very carefully dissect out the entry of the pulmonary veins into the heart. It will probably seem as if the right pulmonary veins and the inferior cava opened into the same portion of the organ, but it will be found subsequently (13. *a.*) that such is not really the case. Note on the exterior of the organ the following points:

a. Its upper flabby *auricular* portion into which the veins open, and its denser lower *ventricular* part.

b. Running around the top of the ventricles is a band of fat, an offshoot of which runs obliquely down the front of the heart, passing to the right of its apex, and indicating externally the position of the internal partition or *septum* which separates the right ventricle, which does not reach the apex of the heart, from the left, which does.

c. Note the fleshy "*auricular appendages*"—one (*left*) appearing below the pulmonary artery; the other (*right*), between the aorta and superior cava.

10. Dissect away very carefully the collection of fat around the origins of the great arterial trunks and that around the base of the ventricles. In the fat will be found—

a. A *coronary artery* arising from the aorta close to the heart, opposite the right border of the pulmonary artery; it gives off a branch which runs in the groove between right auricle and ventricle, and then runs down the dorsal side of the heart on the ventricles.

b. The other *coronary artery*, considerably larger, arises from the aorta dorsal to the pulmonary artery; its main branch runs along the ventral edge of the ventricular septum.

c. The *coronary veins* and *sinus:* small coronary veins will be seen accompanying the arteries; for the coronary sinus see 11. *c.*

11. Open the right ventricle by passing the blade of a scalpel through the heart about an inch from the upper border of the ventricle, and on the right of the band of fat marking externally the limits of the ventricles, and noted above (9. *b.*), and then cut down towards the apex, keeping on the right of this line; cut off the pulmonary artery about an inch above its origin from the heart, and open the right auricle by cutting a bit out of its wall, to the left of the entrances of the venæ cavæ. On raising up by its point the wedge-shaped flap cut from the wall of the ventricle, the cavity of the latter will be exposed.

a. Pass the handle of a scalpel from the ventricle into the auricle;

and also from the ventricle into the pulmonary artery, and make out thoroughly the relations of these openings.

b. Slit open the auricular appendage; note the fleshy projections (*columnæ carneæ*) on its walls, and the smoothness of the rest of the interior of the auricle.   Observe the apertures of the *venæ cavæ*, and note that the pulmonary veins do not open into this auricle.

c. Behind or below the entrance of the inferior cava, note the entrance of the *coronary sinus;* pass a probe through the aperture along the sinus and slit it open; notice the muscular layer covering it in.

12. Raise up by its apex the flap cut out of the ventricular wall, and if necessary prolong the cuts more towards the base of the ven-tricle until the divisions of the *tricuspid valve* come into view.

a. Note the columnæ carneæ on the wall of the ventricle, and the muscular cord (not found in the human heart) stretching across its cavity.   Also the prolongation of the ventricular cavity towards the aperture of the pulmonary artery.

b. Cut away the right auricle, and examine carefully the *tricuspid valve,* composed of three membranous flexible flaps, thinning away towards their free edges; proceeding from near these edges are strong *tendinous cords* (*chordæ tendineæ*), which are attached at their other ends to muscular elevations (*papillary muscles*) of the wall of the ventricle.

c. Slit up the right ventricle until the origin of the pulmonary artery comes into view.   Looking carefully for the flaps of the semi-lunar valves, prolong your cut between two of them so as to open the bit of pulmonary artery still attached to the heart.   Spread out the artery and examine the valves.

d. Each flap makes, with the wall of the artery, a pouch, opposite which the arterial wall is slightly dilated.   The free edge of the valve is turned from the heart, and has in its middle a little nodule (*corpus Arantii*).

13. Open the left ventricle in a manner similar to that employed for the right.   Then open the left auricle by cutting a bit out of its wall above the appendage.   Cut the aorta off about half an inch above its origin from the heart.   The aperture between left auricle and left ventricle can now be examined; also the passage from the ventricle into the aorta, and the entry of the pulmonary veins into the auricle; and the *septum* between the auricles and that between the ventricles.

a. Pass the handle of a scalpel from the ventricle into the auricle; another from the ventricle into the aorta; and pass also probes into the points of entrance of the pulmonary veins.   Observe that no other veins open into this auricle.

b. Slit open the auricular appendage; note the fleshy projections (*columnæ carneæ*) on its interior, and the general smoothness of the rest of the inner wall of the auricle. Notice the *columnæ carneæ* over the inner surface of the ventricular wall, also the considerable thickness of the latter, as compared with that of the right ventricle or of either of the auricles.

c. Carefully raise the wedge-shaped flap of the left ventricle, and cut on towards the base of the heart, until the valve (*mitral*) between auricle and ventricle is brought into view; one of its two flaps will be seen to lie between the auriculo-ventricular opening and the origin of the aorta.

Examine in these flaps their texture, the chordæ tendineæ, the columnæ carneæ, etc., as in the case of the right side of the heart (12).

d. Examine the semilunar valves at the exit of the aorta; then cutting up carefully between two of them, examine the bit of aorta still left attached to the heart, and note the valves more carefully as described in 12. *d.* Note the origins of the coronary arteries in two of the three dilatations (*sinuses of Valsalva*) of the aortic wall above the semilunar flaps.

14. Examine a piece of aorta. Note that when empty it does not collapse; the thickness of its wall; its extensibility in all directions; its elasticity.

15. Compare with the artery the thin-walled flabby veins which open into the heart.

CHAPTER XVI.

THE WORKING OF THE HEART AND BLOOD-VESSELS.

The Beat of the Heart.—It is possible by methods known to physiologists to open the chest of a living animal, such as a rabbit, made insensible by chloroform, and see its heart at work, alternately contracting and diminishing the cavities within it, and relaxing and expanding them. It is then observed that each beat commences at the mouths of the veins which open into the auricles; and from there runs over the rest of the auricles, and then over the ventricles; the auricles beginning to dilate the moment the ventricles start their contraction. Having finished their contraction, the ventricles begin to dilate, and then for some time neither they nor the auricles are contracting, but the whole heart is expanding. The contraction of any part of the heart is known as its *sys'to-le*, and the relaxation as its *di-as'to-le*, and since the two sides of the heart work synchronously, the auricles together and the ventricles together, we may describe a whole "cardiac period" or "heart-beat" as made up successively of *auricular systole, ventricular systole*, and *pause*. In the *pause* the heart, if taken between the finger and thumb feels soft and flabby,

What is seen when the beating heart of a living animal is exposed? When do the auricles begin to dilate? What is the state of the heart for a short time after the end of a ventricular contraction? What is meant by the systole of a part of the heart? What by the diastole? Of what does a cardiac period consist? How does the heart feel to the touch during the pause?

but during the systole it, especially in its ventricular portion, becomes hard and rigid, and diminished in size so as to force blood out of it.

The Cardiac Impulse.—The human heart lies with its apex touching the chest-wall between the fifth and sixth ribs on the left side of the breast-bone. At every beat a sort of tap known as the "cardiac impulse," or "apex beat," may be felt by placing the finger at that point.

Events occurring within the Heart during a Cardiac Period.—Let us commence just after the end of the ventricular systole. At this moment the semilunar valves at the orifices of the aorta and the pulmonary artery are closed so that no blood can flow back from those vessels. The whole heart, however, is soft and distensible, and yields readily to blood flowing into its auricles from the pulmonary veins and the hollow veins; this blood passes on through the open mitral and tricuspid valves, and fills up the dilating ventricles as well as the auricles. As the ventricles fill, back currents are set up along their walls, and carry up the flaps of the auriculo-ventricular valves, so that by the end of the pause they are nearly closed. At this moment the auricles contract; this contraction commences at and narrows the mouths of the veins so that blood cannot easily flow back from the auricles into them; the flabby and dilating ventricles oppose much less resistance, and so the general result is

How during the systole? How is its bulk changed in systole? Where does the apex of the heart touch the chest-wall? What is the cardiac impulse?

What is the position of the semilunar valves just after the end of a ventricular systole? What results from their closure? In what condition is the heart in general? What parts of it does blood enter? From what vessels? What cavities does this blood fill? What happens as the ventricles fill? What is the position of the valves at the end of the pause? Where does the auricular contraction commence? What is the main result of the auricular contraction?

that the contracting auricles send blood mainly into the ventricles and hardly any back into the veins. The increased current into the ventricles produces a greater back current on the sides, which, as the auricles cease their contraction, and the filled ventricles become tense and press on the blood inside them, completely closes the auriculo-ventricular valves.

The auricular contraction now ceases, and the ventricular begins. The blood in each ventricle is imprisoned between the auriculo-ventricular valves behind and the semilunar valves in front. The former cannot yield on account of the chordæ tendineæ fixed to their edges; the semilunar valves, on the other hand, can open outwards from the ventricle and let the blood pass on; but they are kept tightly shut by the pressure of the blood in the aorta and pulmonary artery, just as the lock-gates of a canal are by the pressure of the water on them. In order to open the canal-gates water is let in or out of the lock until it stands at the same level on each side of them; but they might be forced open without this by applying sufficient power to overcome the higher water pressure on one side. It is in this latter way that the semilunar valves are opened.

The contracting ventricle tightens its grip on the blood inside it. As it squeezes harder and harder, at last the pressure on the blood in it becomes greater than the pressure exerted on the other side of the valves by the blood in

What is the consequence of the increased flow into the ventricles due to the auricular contraction?

What happens when the ventricle begins to contract? Why cannot the imprisoned blood escape back into the auricle? How are the semilunar valves kept closed? Illustrate. How might we force open the gates of a canal lock without bringing the water to the same level on each side?

How are the semilunar valves opened?

the arteries, the valves are forced open, and the blood begins to pass out; the ventricle continues to contract until it has obliterated its cavity and completely emptied itself. Then it commences to relax, and blood to flow back into it from the arteries. This back current, however, catches the pockets of the semilunar valves, drives them back, and closes the valve so as to form an impassable barrier, and so the blood which has been forced out of the ventricle is hindered from flowing directly back into it.

Use of the Papillary Muscles.—In order that the contracting ventricles may not force blood back into the auricles, it is essential that the flaps of the mitral and tricuspid valves be held together across the openings which they close, and not pushed back into the auricles. If they were like swinging doors and opened both ways they would be useless ; they must so far resemble an ordinary door as only to open in one direction, namely, from the auricle to the ventricle. At the commencement of the ventricular systole this is provided for by the chordæ tendineæ, which are of such a length and so arranged as to keep the valve flaps shut across the opening, and to maintain their edges in contact. But, as the contracting ventricles shorten, the chordæ tendineæ would be slackened and the valve-flaps pushed up into the auricle. The little papillary muscles prevent this. Shortening as the ventricular systole proceeds, they keep the chordæ taut and the valves closed.

What then happens? How long does the ventricle continue to contract? What then follows? How are the semilunar valves closed? What is essential in order that blood may not be forced back from ventricle to auricle? Illustrate. How is the pushing back of the valve-flaps between auricle and ventricle prevented at the beginning of a ventricular systole? When would the chordæ tendineæ be slackened? What would result? How is the slackening prevented?

Sounds of the Heart.—If the ear be placed on the chest of another person over the heart region, two distinguishable sounds will be heard during each round of the heart's work. They are known respectively as the *first* and *second sounds* of the heart. The first is of lower pitch and lasts longer than the second and sharper sound; vocally their character may be tolerably imitated by the syllables *lŭb, dŭp.* The cause of the second sound is the closure, or, as one might say, the "clicking up" of the semilunar valves. The first sound takes place during the ventricular systole, and is probably due to vibrations of the tense ventricular wall at that time. In many forms of heart disease these sounds are modified or cloaked by additional sounds which arise when the cardiac orifices are roughened, or narrowed, or dilated, or the valves inefficient. A physician often gets important information as to the nature of a heart disease by studying these new or altered sounds.

Function of the Auricles.—The ventricles have to do the work of pumping the blood through the blood-vessels. Accordingly their walls are far thicker and more muscular than those of the auricles; and the left ventricle, which has to force the blood over most of the body, is stouter than the right, which has only to send blood around the comparatively short pulmonary circuit. The circulation of the blood is, in fact, maintained by the ventricles, and we have to inquire what is the use of the auricles. Not unfre-

What do we hear on listening over the heart region of a living person's chest? What are the sounds called? How does the first differ from the second? What words give some idea of their character? What is the origin of the second sound? Of the first? What occurs as regards the heart sounds in many forms of heart disease?

What work have the ventricles to do? How do their walls differ from those of the auricles? Which ventricle has the thicker wall? Why? What part of the heart maintains the blood flow?

quently the heart's action is described as if the auricles first filled with blood, and then contracted and filled the ventricles; and then the ventricles contracted and drove the blood into the arteries. From the account given above, however, it will be seen that the events are not accurately so represented, but that during all the pause the blood flows on through the auricles into the ventricles, which latter are already nearly full when the auricles contract; this contraction merely completes the filling of the ventricles, and finishes the closure of the auriculo-ventricular valves. The main use of the auricles is to afford a reservoir into which the veins may empty while the comparatively long-lasting ventricular contraction is taking place.

The Work done daily by the Heart.—At each beat each ventricle pumps on rather more than six ounces (say fourteen tablespoonfuls) of blood. The elastic aorta and the pulmonary artery are full, and resist the pumping of more liquid into them, just as an elastic bag filled with water could only have more sent into it by force; to get more in one would have to stretch the bag more. The resistance opposed by these arteries to receiving blood from the heart has been measured in some of the lower animals, and calculations made from them to man. According to these the work which the left ventricle does every day, sending 6¼ ounces of blood seventy times a minute into the aorta, is enough to lift one pound 325,584 feet high; and the work done by the right ventricle would lift one pound 108,528

What parts of the heart does the blood enter during the pause? What is the condition of the ventricles as regards fullness at the end of the pause? What is done by the auricular contraction? What is the chief use of the auricles?

How much blood does each ventricle pump out at every beat? What resists the ventricular emptying? Illustrate. How much work does a man's left ventricle do daily? How much the right?

feet. The work done daily by the ventricles of the heart together is equal to that required to raise one pound 434,112 feet from the earth's surface, or, what comes to the same thing, more than 193 tons one foot high.

If a man weighing 165 pounds climbed up a mountain 2644 feet high the muscles of his legs would probably be greatly tired at the end of his journey, and yet in lifting his body that height they would only have done as much work as his heart does every day without fatigue in pumping his blood.

No doubt the fact that more than half of every round of the heart's activity is taken up by the pause during which its muscles are relaxed and its cavities filling with blood, has a great deal to do with the patient and tireless manner in which it pumps along, minute after minute, hour after hour, and day after day, from birth to death.

The Pulse.—When the left ventricle of the heart contracts it forces on about six ounces of blood into the aorta, which, with its branches, is already quite full of blood. The elastic arteries are consequently stretched by the extra blood, and the finger laid on one feels it dilating; this dilatation of an artery following each beat of the heart is called *the pulse;* it is easiest felt on arteries which lie near the surface of the body, as *the radial artery,* near the wrist, and *the temporal artery,* on the brow.

The arteries at their ends furthest from the heart lead

How much both ventricles together?

How high would a man have to climb in order to do as much work by the muscles of his legs as the heart does in a day?

How may we account for the fact that the heart does not become fatigued and unable to work?

What happens when the left ventricle of the heart contracts? What results in the arteries? What is the pulse? Name arteries on which the pulse is easily felt.

into capillaries; before the next heart-beat occurs they pass on into these minute vessels as much blood as the aorta received during the preceding ventricular systole; consequently they shrink again during the pause, just as a piece of rubber tubing with a small hole in it, when overfilled with water, would gradually collapse as the water flowed out of it. The next beat of the heart again overfills and expands the arteries, and so on; at each heart-beat there is a dilatation of the arteries due to the blood sent into them from the ventricle, and between each beat there is a partial collapse of the arteries, due to their emptying blood into the capillaries.

What may be learnt from the Pulse.—The pulse being dependent on the heart's systole, "feeling the pulse" of course primarily gives a convenient means of counting the rate of beat of that organ. To the skilled touch, however, it may tell a great deal more; as, for example, whether it is a readily compressible or "soft pulse," showing that the heart is not keeping the arteries properly filled up with blood, or tense and rigid ("a hard pulse"), indicating that the heart is keeping the arteries excessively filled, and is working too violently, and so on. In healthy adults the pulse rate may vary from sixty-five to seventy-five a minute, the most common rate being seventy-two. In the same individual it is faster when standing than when sitting, and when sitting than when lying down. Any exercise in-

Into what do the final arterial branches open? How much blood is sent into the capillaries during a cardiac period? What change takes place in the bulk of the arteries during the interval between two ventricular contractions? Illustrate. What happens in the arteries during each heart-beat? Why? What during each heart pause? Why?

How may we conveniently count the rate of heart-beat? What does a soft pulse indicate? A hard pulse? What is the most common pulse rate in health? Within what limits may it vary? How is it influenced by the position of the body?

creases its rate temporarily and so does excitement; a sick person's pulse should not therefore be felt when he is nervous or excited (as the physician knows when he tries first to get his patient calm and confident). In children the pulse is quicker than in adults, and in old age slower than in middle life.

The Flow of the Blood in the Capillaries and Veins.—The blood leaves the heart intermittently and not in a regular stream, a quantity being forced out at each systole of the ventricles; before it reaches the capillaries, however, this rhythmic movement is transformed into a steady flow, as may readily be seen by examining with a microscope thin transparent parts of various animals, as the web of a frog's foot, a bat's wing, or the tail of a small fish. In consequence of the steadiness with which the capillaries supply the veins the flow in these latter is also unaffected directly by each beat of the heart; if a vein be cut the blood wells out uniformly, while a cut artery spurts out with much more force, and in jets which are more powerful at regular intervals corresponding with the contractions of the ventricles.

The Circulation of the Blood as seen in the Frog's Web.—There is no more fascinating or instructive spectacle than the circulation of the blood as seen with the microscope in the thin membrane between the toes of a frog's hind limb. Upon focusing beneath the outer layer of the skin a network of minute arteries, veins, and capillaries,

How by exercise? Why should an invalid's pulse not be felt when he is excited? How does age affect the pulse rate?

In what manner does the blood leave the heart? How is its flow altered before reaching the capillaries? How may this be observed? Is the flow in the veins rhythmic or steady? Why? How does the bleeding from a cut artery differ from that of a cut vein?

What comes into view on examining a frog's web with a microscope?

with the blood flowing through them, comes into view (Fig. 61). The arteries, *a*, are readily recognized by the fact that the flow in them is fastest and from larger to smaller branches. The smallest are seen to end in capillaries, which form networks, the channels of which are all nearly equal in size. In the veins arising from the capillaries the flow is from smaller to larger trunks, and slower than in the arteries, but faster than in the capillaries.

Why the Blood flows slowest in the Capillaries.—The reason of the slower flow of the capillaries is that their united area is considerably greater than that of the arteries supplying them, so that the same quantity of blood flowing through them in a given time has a wider channel to flow in and therefore moves more slowly. The area of the veins is smaller than that of all the capillaries, but greater than that of the arteries, and so the rate of movement in them is intermediate.

We may picture to ourselves the vascular system as a double cone, widening from the ventricles to the capillaries, and narrowing from the latter to the auricles. Just as water forced in at a narrow end of this would flow quickest there, slowest at the widest part, and quicker again where it passed out the other narrow end, so the blood flows quick in the aorta and hollow veins,* and slow in the capillaries,

How may the arteries be recognized? In what are the smallest arteries seen to end? Do the capillaries vary much in size? What is the direction of flow in the veins? How does its rate differ from that in the arteries? From that in the capillaries?

Why does blood flow slowest through the capillaries? Why in the veins quicker than in the capillaries, but slower than in the arteries?

How may we picture the vascular system? Illustrate. How do capillaries differ in size from the large arteries?

*A good illustration taken from physical geography is afforded by the Lake of Geneva, in Switzerland. This is supplied at one end by a river which derives its water from the melting glaciers of some of the Alps. From its other end the

which, though thousands of times smaller than the great arteries and veins, are millions of times more numerous. The channel through which the blood flows in them is, therefore, when they are all taken together, very much greater than that to which it is confined in the large arterial and venous trunks.

Why there is no Pulse in the Capillaries and Veins.— The heart sends blood into the arteries not steadily but intermittently; each beat forces in some blood, and then comes a pause before the next beat. Accordingly the flow in the larger arteries is not even and continuous, but jerky, as indicated by the pulse.

But in the capillaries the flow is quite steady, and yet the capillaries are supplied by the smaller arteries. We have to inquire how this is brought about.

The disappearance of the pulse is due to two things, (1) the fact that in the tiny capillaries the blood meets with considerable resistance to its flow, dependent on friction, and (2) that the arteries are very elastic.

On account of friction in the capillaries the arteries have difficulty in passing on blood through them; blood therefore accumulates in the aorta and its large branches and stretches their elastic walls. The stretched arteries press all the time on the blood inside them, and constantly keep squeezing it on into the small arteries and the capillaries;

How in number? Is the total blood channel greater in arteries or capillaries? In veins or capillaries?

Why is the blood-flow in the great arteries not steady? Name vessels in which it is steady.

To what is the loss of pulse in the capillaries due? What results from friction in the capillaries? What is done by the stretched arteries?

---

water is carried off by the river Rhone. In the comparatively narrow inflowing and outflowing rivers the current is rapid; in the wide bed of the lake it is much slower.

both while the heart is contracting and between two heart-beats. The heart, in fact, keeps the big elastic arteries over-distended with blood; before they have had time to nearly empty, another systole occurs and fills them up tight again; so all the while the walls of the arteries are stretched and keep pressing on the blood inside them, and steadily forcing it on into the capillaries. The heart keeps the arteries over-full, and the stretched elastic arteries drive the blood through the capillaries. As the arteries are always stretched and always pressing on the blood the capillaries receive a steady supply, and the flow through them is uniform. This even capillary flow passes on a steady blood stream to the veins.*

The object of having no pulse in the capillaries is to diminish the danger of their rupture. As we have seen, materials from the blood have to ooze through their walls to nourish the organs of the body, and wastes from the organs to soak back into the blood that they may be carried off. Their walls have therefore to be very thin; and if the

When? What does the heart do? What happens before the arteries have had time to empty? What is the condition of the arterial walls all through life? What results from their stretched condition? What keeps the arteries tightly filled? What sends blood through the capillaries? How do the capillaries get a steady blood supply? Why do we find a uniform current in the veins?

What is gained by having no pulse in the capillaries? What must food materials in the blood do before they can nourish the body? What must the wastes of the organs do? Why must the capillary walls be very thin?

---

* "Every inch of the arterial system may, in fact, be considered as converting a small fraction of the heart's jerk into a steady pressure, and when all these fractions are summed up together in the total length of the arterial system no trace of the jerk is left. As the effect of each systole becomes diminished in the smaller vessels by the causes above mentioned, that of this constant pressure becomes more obvious, and gives rise to a steady passage of the fluid from the arteries towards the veins. In this way, in fact, the arteries perform the same functions as the air-reservoir of a fire-engine, which converts the jerking impulse given by the pumps into the steady flow of the delivery hose."— HUXLEY.

blood were sent into them in sudden jets at each beat of the heart, they would run much risk of being torn.

The Muscles of the Arteries.—The arteries have rings of plain muscular fibre in their walls; when these contract they narrow the artery, and when they relax they allow it to widen under the pressure of the blood in its interior. The vessel then carries more blood to the capillaries of the organ which it supplies. *Blushing* is due to a relaxation of the muscular layer of the arteries of the face and neck, allowing more blood to flow to the skin.

Why the Arteries have Muscles.—The amount of blood in the body is not sufficient to allow of a full stream of blood through all its organs at one time: the muscular fibres controlling the diameter of the arteries are used to regulate the blood-flow in such a manner that parts hard at work shall get an abundant supply, and parts at rest shall only get just enough to keep them nourished. Usually when one set of organs is at work and its arteries dilated, others are at rest and their arteries contracted. Few persons, for example, feel inclined to do brain-work after a heavy meal; for then a great part of the blood of the whole body is led off into the dilated vessels of the digestive organs, and the brain gets but a small supply. On the other hand, when the brain is at work its vessels are dilated, and often the whole head flushed; and when the muscles are exercised, a great portion of the blood of the body is carried off to them; there-

What would be apt to happen if blood were sent into them in sudden jets?

How are the muscles of the arteries arranged? What results from their contraction? From their relaxation? To what is blushing due?

Why cannot all the organs have a full blood stream through them at the same time? For what purpose are the muscular fibres in the walls of arteries used? What is the usual condition of the arteries of a resting organ?

fore, hard thought or violent exercise soon after a meal is very apt to produce an attack of indigestion by diverting the blood from the abdominal organs, where it ought to be at that time. Young persons whose organs have a super-abundance of energy, enabling them to work under unfavorable conditions, are less apt to suffer in such ways than their elders. One sees boys running actively about after eating, when older people feel a desire to sit quiet or even to go to sleep.

Taking Cold.—When the skin is chilled its arteries contract, as shown by the pallor of the surface. This throws an undue amount of blood into internal parts, whose vessels become gorged with blood or "congested," and congestion very easily passes into inflammation. Consequently, prolonged exposure of the surface to cold is very apt to be followed by inflammation of parts inside the body, and give rise to a so-called "cold" (which is really an inflammation) of the mucous membranes of the head, or throat, or lungs; or of the intestines, causing diarrhœa. In fact, the common summer diarrhœa is far more often due to a chill of the surface leading to intestinal inflammation than to the fruits eaten in that season, which are so often blamed for it. The best preventive is to wear when exposed to sudden changes of temperature, a woollen or at least a cotton garment over the trunk of the

Why is it not wise to take hard exercise or do severe mental work soon after eating? Why do young persons suffer less from exercise soon after dinner than do their elders?

What happens to its arteries when the skin is chilled? How does this manifest itself? What is its result on the blood-supply of internal parts? What is congestion? Into what diseased state does it often pass?

What diseases are apt to follow a surface chill? What is the most frequent cause of summer diarrhœa? What should be worn when liable to exposure to considerable changes of temperature?

body; linen permits any change in the external temperature to act almost at once upon the skin. After an unavoidable exposure to cold or wet, the thing to be done is
of course to maintain the cutaneous circulation; movement
should be persisted in, and a thick dry outer covering put
on until warm and dry underclothing can be obtained.

In healthy persons, a temporary exposure to cold, as
a plunge in a bath, is good, since in them the sudden contraction of the cutaneous arteries soon passes off, and is
succeeded by a dilatation causing a warm healthy glow on
the surface. If the bather remain too long in cold water,
however, this reaction passes off, and is succeeded by a more
persistent chilliness of the surface, which may last all day.
The bath should therefore be left before this occurs; but
no absolute time can be stated, as the reaction is more
marked and lasts longer in strong persons and in those used
to cold bathing than in others.

By partially paralyzing certain nerves of the heart
which keep its beating in check, wines and spirits quicken
the beat of the heart, leaving it less time for repair between
its strokes. This causes the heart's walls to be at first thick-
ened (*hypertrophied*) and then its cavities dilated by the excessive work (p. 111) which alcoholic drinks cause it to perform. If, as is usually the case, fatty degeneration ensues,
the organ gradually becomes too feeble to pump the blood
around the body, and death results. A heart which has
undergone this change is commonly spoken of by patholo-
gists as a "whiskey heart;" for although fatty degenera-

Why does linen not form a good inner garment under such circumstances? What should be done after unavoidable exposure to
cold or wet? How does alcohol injure the heart?

Why is a plunge in a cold bath useful to healthy persons? What
results if a person remains too long in cold water? When should a
cold bath be left? What persons may remain longest with safety in
a cold bath?

tion of the heart may occur from other causes, alcoholic indulgence is the most frequent one. Fatty liver or fatty heart is rarely if ever curable ; either will ultimately cause death.

APPENDIX TO CHAPTER XVI.

1. A frog may be used to illustrate the beat of the heart. Anatomically a frog's heart differs in many respects from that of a mammal, but the phenomena of systole and diastole are essentially the same.

2. Etherize a frog as before described. Cut off its head. Check bleeding and destroy its spinal cord by forcing a pointed wooden peg along the spinal canal.

3. Laying the animal on its back, carefully divide with scissors the skin along the middle line of the ventral surface for its whole length. Make cross cuts at each end of this longitudinal one and pin out the flaps of skin.

4. Next pick up with forceps the remaining tissues of the ventral wall near its posterior end, and carefully divide them longitudinally a little on the left side of the middle line; being very careful not to injure either the viscera in the cavity beneath or a large vein (*anterior abdominal*) running along the wall in the middle line.

5. About the point where you see this vein passing from the wall to enter among the viscera of the ventral cavity, you will come to the bony and cartilaginous tissues of the sternal region. Raise the posterior cartilage in your forceps, make a short transverse cut in front of the vein, and, looking beneath the sternum, note the pericardium with the heart beating inside it. Divide the fibrous bands which pass from the pericardium to the sternum, and with scissors cut away sternum, etc., taking great care not to injure the heart.

6. Push a rod about half an inch in diameter down the animal's throat so as to stretch the parts, and then picking up the pericardium in a pair of forceps, open it and gently cut it away from about the heart; push aside any lobes of the liver which lie on the latter organ. In the heart thus exposed note—

a. Its *beat;* a regularly alternating contraction (*systole*) and dilatation (*diastole*).

b. In consequence of the destruction of the spinal cord comparatively little blood now flows through the heart, but during the contraction you will be able to observe that the ventricular portion, which will be readily recognized, becomes paler; and during diastole again becomes deeply colored, getting more or less filled up with blood which shows through its walls.

c. Observe that each contraction starts at the auricular end and travels towards the ventricular; this may be more easily seen by-and-by, when the heart begins to beat more slowly.

7. The specimen may be put aside under a bell-jar with a wet sponge, or a piece of flannel soaked in water. If kept from drying the heart will go on beating for hours.

8. To demonstrate the action of the valves of the heart, obtain two uninjured sheep's hearts from a butcher. Remove them from the pericardium, taking care not to injure the vessels.

9. Cut off the apex of one heart so as to open the ventricles. Then fill up the stumps of the aorta and the pulmonary artery with water. As the water is poured in the semilunar valves will be seen to close up and block the passage to the ventricle, so that the stump of the vessel remains full for some time. The valves rarely act quite perfectly in a heart removed from the body and treated as above, but they will support the water column quite long enough to illustrate their action.

10. Carefully cut the auricles away from the other sheep's heart, taking great care not to injure the ventricles or the auriculo-ventricular valves. Then holding the ventricles, apex down, in one hand, pour water in a stream into them from a pitcher held about a foot above them. As the ventricles fill, the flaps of the mitral and tricuspid valves will be seen to float up and close the auriculo-ventricular orifice, illustrating their movement as the ventricle fills during its diastole in the natural working of the heart.

11. The manner in which the elasticity of the arteries and the friction resistance to flow in the capillaries together serve to turn a rhythmic into a steady flow may be readily demonstrated as follows:

Take an elastic bag such as is commonly sold with enema apparatus in drug stores, and having an entry and exit tube provided with valves In the exit tube place a piece of glass tubing six feet long. Put the entry tube of the bag in a basin of water. On pumping, an intermittent flow of water, corresponding to the strokes of the pump, will be obtained from the glass tube. Connect a very fine glass nozzle with the end of the long tube; on pumping, less water can be forced through, and the outflow is still rhythmic.

12. Replace the glass tube by a rubber tube of the black, highly elastic kind: on pumping we get again a rhythmic outflow. Now connect your narrow nozzle to the end of the rubber tube, and pump: the outflow will be nearly constant, because the rubber tube not being able to empty itself as fast as the water is pumped into it, becomes

stretched, and in the interval between two strokes of the pump it keeps on squeezing out the extra water accumulated in it. The longer and more elastic the tube, the quicker and stronger the stroke of the pump, and the narrower the exit, the more steady will be the outflow. In the body the heart keeps the arteries very tightly stretched all the time, and they keep up accordingly a steady flow into the capillaries. The experiment shows that to get such a steady flow two things are necessary : (1) that the tubes fed directly by the heart shall be highly elastic, and (2) that there shall be considerable resistance to the exit from their outflow ends

CHAPTER XVII

THE OBJECT AND THE MECHANICS OF RESPIRATION.

The Object of Respiration.—Blood is renewed, so far as ordinary food materials are concerned, by substances either directly absorbed by the blood-vessels of the alimentary canal, or taken up by the lymphatics of the digestive tract and afterwards poured into the blood. But in order that energy may be set free for use by the tissues of the body (Chap. IX.), oxidations must occur, and the continuance of these vital oxidations depends on a constant supply of oxygen. As their result, waste substances are produced, which are no longer of use to the body, but detrimental to it if present in large quantity. The most abundant of these wastes is carbon dioxide gas.

The function of respiration has for its objects (1) to renew the supply of oxygen in the blood, and (2) to get rid of the carbon dioxide produced in the different organs.

The Respiratory Apparatus.—This consists primarily of two elastic bags, *the lungs*, placed in the thorax, filled with air, and communicating by *the air-passages* with the sur-

How is the blood renewed as regards ordinary food matters? What must occur that energy be set free for use by the body? What is necessary that the oxidations may continue? What do the oxidations produce? Which is the most abundant waste substance of the body?

What are the objects of respiration?

Of what does the respiratory apparatus primarily consist?

rounding atmosphere. In the lungs the pulmonary capillary blood-vessels form a very close network: through their walls the blood gives off to the air in the lungs carbon dioxide, and takes from this air oxygen. The air in the lungs consequently needs renewal from time to time: otherwise it would no longer have oxygen to give to the blood, and would become so loaded with carbon dioxide as to no longer take that waste product from it. This renewal is effected by the working of a system of muscles, bones, and cartilages whose co-operation brings about that alternating expansion and contraction of the chest which we call *breathing.* When the chest contracts, air deprived of its oxygen and polluted with wastes is expelled from the lungs; and when it expands, fresh air, rich in oxygen, and containing hardly any carbon dioxide, is taken into them.

The respiratory organs are, therefore, (1) the lungs; (2) the air-passages; (3) the vessels of the pulmonary circulation, including the pulmonary artery bringing the blood to the lungs, the pulmonary capillaries carrying it through them, and the pulmonary veins conveying it from them; (4) the muscles, bones, and gristles which are concerned in producing the breathing movements.*

The Air-Passages.—Air reaches the pharynx through the nose or mouth (Fig. 1): on the ventral side of the pharynx

What vessels form a close network in the lungs? What takes place through the walls of these vessels? Why must the air in the lungs be renewed? How is the renewal brought about? What is breathing? What happens when the chest contracts? What when it expands ?
Enumerate the respiratory organs.
Through what passages does air reach the pharynx?

---

* To these should be added (5) the nerve centres and nerves which contro; the muscles of respiration, and which will be subsequently considered (see Chap. XX).

(Fig. 41) is an aperture through which it passes into the *larynx* or voice-box (*a*, Fig. 64), which lies in the upper part of the neck. From the larynx air passes on through the *windpipe* or *trachea;* this enters the chest, in the upper

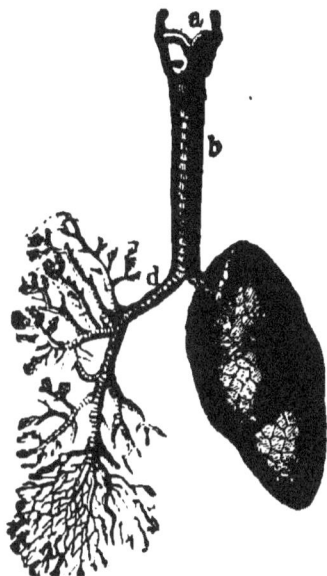

Fig. 64.   Fig. 65.

Fig. 64.—The lungs and air-passages seen from the front. On the left of the figure the pulmonary tissue has been dissected away to show the ramifications of the bronchial tubes. *a*, larynx ; *b*, trachea; *d*, right bronchus. The left bronchus is seen entering the *root* of its lung.

Fig. 65.—A small bronchial tube, *a*, dividing into its terminal branches, *c*; these have pouched or sacculated walls and end in the sacculated alveoli, *b*.

part of which it divides into a *right* and a *left bronchus.* Each bronchus enters a lung, and divides in it into a vast number of very small tubes, called the *bronchial tubes.* The last and smallest bronchial tubes (*a*, Fig. 65) open into subdivided elastic sacs, *b, c,* with pouched walls.

What aperture is found on the ventral side of the pharynx? Where does the larynx lie? Where does the air go from the larynx? Into what does the trachea divide? Where? What are the bronchial tubes? How do the final bronchial tubes terminate?

Structure of the Windpipe and its Branches. — The trachea, bronchi, and bronchial tubes are lined by mucous membrane, outside of which is a supporting stratum composed of connective and plain muscular tissues. Their walls also contain cartilaginous rings or half-rings which keep them open. Below the projection on the throat known as Adam's apple (due to the larynx, see Chap. XXIV.) there may readily be felt in thin persons the stiff windpipe passing down to the top of the chest.

The Cilia of the Air-Passages. — The mucous membrane of the trachea and its branches, down to almost the smallest, has a layer of *ciliated cells* on its surface. Each of these cells has on its end turned towards the cavity of the tube a tuft of from twenty to thirty slender threads which are in constant motion; they lash forcibly towards the throat, move gently back again, and then once more violently towards the outlet of the air-passage. These moving threads are called *cilia*. Swaying in the mucus secreted by the membrane which they line, they sweep it on to the throat, where it is coughed or "hawked" up.

Imagine a man rowing in a boat at anchor. The sweep of the oars will drive the water back and not the boat forwards. So these little oars, the cilia, being anchored on the mucous membrane drive on the secretion which bathes its surface.

With what are the windpipe and its branches lined? What lies outside this lining? What do these walls also contain? What is the use of the cartilages? What is "Adam's apple"? What may be felt below it in front of the neck?

What lines the mucous membrane of the windpipe and its subdivisions? What does each ciliated cell bear on its free end? How do the threads move? What are they called? What is the use of the cilia of the air-passages?

Illustrate how they push on the liquid they move in.

Bronchitis, or "a cold on the chest," is an inflammation of the membrane lining the bronchial tubes, in consequence of which it swells, and secretes an extra amount of mucus. The swelling and secretion tend to close the tubes and interfere with the free passage of air in breathing.

The Lungs consist of the bronchial tubes and their terminal dilatations, together with blood-vessels, lymphatics, and nerves, all bound firmly together by elastic tissue. The expansions called *"air-cells"* * at the end of each final branch of a bronchial tube (Fig. 66) are relatively very large, and their surface is still further increased by the pouches (*a, b*) which project from them. Their walls are highly elastic, and contain a close network of capillary blood-vessels, supplied by the pulmonary artery and emptying into the pulmonary veins. Through the extremely thin lining of the air-cell, and the thin wall of the capillaries imbedded in it, oxygen is absorbed by the blood from the air in the air-cell and carbon dioxide given up to it. It

Fig. 66.—Two alveoli of the lung highly magnified. *b, b,* the *air-cells,* or hollow protrusions of the alveolus, opening into its central cavity; *c,* terminal branches of bronchial tube.

What is bronchitis? How does an attack of bronchitis interfere with breathing?

Of what do the lungs consist? What are the air-cells? How is their surface increased? What do their walls contain? By what are their capillaries filled, and into what do they empty? What interchange between blood and air occurs in the air-cells of the lungs?

---

* *Cell* is here used in its primitive sense of a small chamber or cavity (*Latin,* cellula), and not in its modern histological signification, as one of the distinct pieces of living matter recognized by the microscope as serving to build up the body by their accumulation and co-operation.

has been calculated that if the walls of the air-cells were all spread out flat and placed side by side they would cover an area of 2600 square feet. This great surface, therefore, represents the area of the body by which oxygen is received and carbon dioxide given off.

The Pleura.—The exterior of each lung, except where its bronchus and blood-vessels enter it, is covered by a thin elastic serous membrane, *the pleura* (Fig. 2). This membrane also lines the inside of the chest. Its surface in health is kept moistened by a small quantity of lymph. In consequence of its smoothness and moisture, during the breathing movements the chest wall and lung glide over each other with hardly any friction.

Pleurisy is inflammation of the pleura. In its early stages it is usually associated with sharp pain on drawing the breath. Later on a large quantity of lymph is often poured out by the inflamed pleura, filling up the cavity which should be occupied by the lung, and pressing the latter up into a small mass, very inefficient for breathing purposes.

The Elasticity of the Lungs.—The lungs are so elastic as to be like a thin india-rubber bag. If we tie a tube tightly into a bronchus and blow in air the lung will dilate, but as soon as we cease blowing and leave the tube open, it will shrink up again. Yet in the chest the lungs always remain so expanded as to completely fill up all the space left for them by the heart and other things contained in the thorax. How is this?

How large is the surface of the body set apart for oxygen reception and carbon dioxide elimination?

What is the pleura? What does it cover besides the outer surface of each lung? What is its condition in health? What is its use?

What is pleurisy? What symptom usually accompanies its early stages? What happens later?

What do the lungs resemble in their elasticity? How may this elasticity be demonstrated? What is the natural condition of the lungs in the chest?

Why the Lungs remain expanded.—We may best understand this by considering a thin india-rubber bag, such as a
toy balloon. If the neck of the bag is open it collapses.
Why? Because the atmosphere, which pushes on all
things near the earth's surface with a pressure of about
15 lbs. on each square inch (1033 grams on each square cen-
timetre), presses equally on the outside and the inside of
the bag. These pressures balance one another, and the bag
collapses on account of its elastic contractility.

We can expand such a bag in two ways. We may blow
air into it forcibly, and so make the pressure inside sufficiently greater than the opposing aërial pressure outside to
overcome the elasticity of the bag. But we can also dis-
tend the bag, not by increasing the aërial pressure inside it,
but by diminishing that outside it.

Suppose (Fig. 67) we tie our rubber bag (d) on the lower
end of the tube b, which, like the tube c, passes
air-tight through a cork, and then fit the cork
tightly into the glass bottle A. The air will
then press with its full weight on the inside of
the bag through b and on the outside of it
through c, and the bag will remain collapsed.
If now we suck air out through c we diminish
its pressure on the outside of the bag without
altering the atmospheric pressure on its inside:
d will therefore begin to expand, because the
pressure inside it is no longer counterbalanced by the pressure outside. The more air we suck out of c (that is, the more
we diminish the atmospheric pressure on the exterior of d)

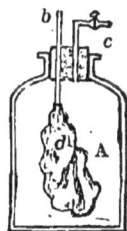

FIG. 67.—Dia-
gram illustrat-
ing the pressure
relationships of
the lungs in the
thorax.

Why does a thin rubber bag collapse when its neck is open?
How can we expand such an elastic bag? Describe a model illustrating the means by which the lungs are kept expanded in the chest,
and explain its action?

the more will the bag expand. Finally, if the rubber bag is distensible enough, when all the air in the bottle is sucked out d will be distended by the push of the air inside it until it completely fills the bottle, whose walls prevent it from going any farther. If now we open c and let in the outside air, the bag will again collapse to its original shrunken dimensions.

Application to the Lungs.—The above experiment illustrates very perfectly how the lungs are kept distended during life. The chest is an air-tight chamber containing, among other things, the elastic hollow lungs. On the interior of the lungs the weight of the atmosphere presses through the air-passages which lead to them, and answer to the tube b in Fig. 67. Outside the lungs in the thorax there is no air at all, and the uncounterpoised aërial pressure inside them overcomes their tendency to shrink, and expands them so as to completely fill all holes and corners of the thoracic chamber not occupied by other organs. If, however, we make a hole in the chest wall and let the air press on the outside of the lungs they collapse at once.

How the Air is renewed in the Lungs.—Suppose in Fig. 67 that the bottle A has a movable bottom, by pulling down which its capacity can be increased, and that all air has been sucked out of the bottle through c, which is then closed. The elastic bag will then be distended by the weight of the atmosphere acting upon its interior so as to fill the bottle and press against its sides. If now the movable bottom of A be pulled down so as to make the cavity of the

Apply to the lungs the facts illustrated by the model represented in figure 67.

Suppose we have a bottle like that in Fig. 67, but with a movable bottom, what would happen when the bottom was pulled down ?

bottle larger, the bag would have a larger space to fill. The impediment on its outside being removed the atmospheric pressure on its inside would expand it still further, and the elastic bag would swell out to fill the extra space. As the movable bottom was pulled down, the bag would enlarge and receive fresh air

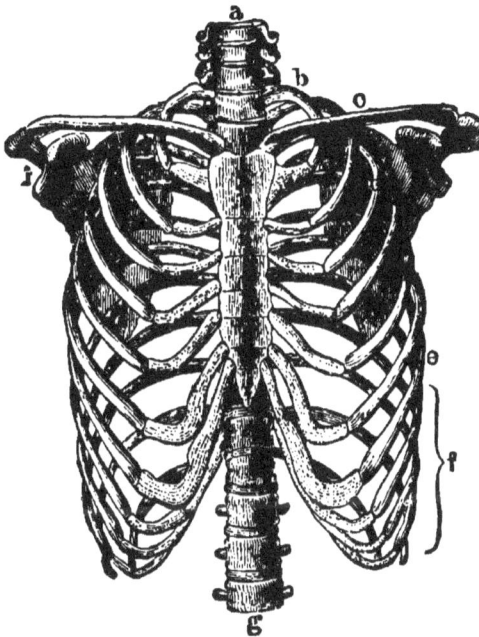

· Fig. 68.—The skeleton of the thorax. *a, g,* vertebral column; *b,* first rib; *c,* clavicle; *d,* third rib; *i,* glenoid fossa.

through *b.* When the bottom was raised again the bag would diminish, and some air be driven out of it through *b.* It is in quite a similar way that the air is renewed in our lungs by breathing. When we breathe-in, the thoracic cavity is enlarged and air enters the lungs; when we

How would the bag inside it be affected?
What would the bag receive? What would happen when the bottom was again raised? What happens when we breathe-in?

breathe-out, the thorax is diminished and air driven out of the lungs.

Inspiration and Expiration.—The process of taking air into the lungs is known as *inspiration*, that of expelling it as *expiration*. On the average, fifteen to eighteen inspirations and expirations occur in each minute. We therefore breathe in and out about once for every four beats of the heart.

The Structure of the Thorax.—The thorax is a conical cavity with a rigid supporting skeleton (Fig. 68) formed by the dorsal vertebræ behind, the breast bone in front, and the ribs and rib cartilages on the sides. Between and over these lie muscles, and the whole is covered air-tight by the skin outside and the pleura (p. 266) inside. Above, it is closed by the muscles and other organs of the neck; and below by a movable bottom, the diaphragm. The air-tight chamber thus bounded can be enlarged in all three diameters, but especially in the vertical, and in that running from the spinal column to the breastbone (dorso-ventral diameter).

The Vertical Enlargement of the Thorax.—This is brought about by the contraction of the diaphragm, which (as may be seen in Fig. 69), is a thin sheet of muscle, with a fibrous membrane in its centre serving as a tendon. In rest the diaphragm is dome-shaped, its concavity being turned towards the abdomen. From the tendon on the crown of

What when we breathe-out?

What is inspiration? Expiration? How often does each occur in a minute? Compare the rate of respiration and of the heart-beat?

What is the form of the thorax? What forms its skeleton? What lie between and over its bones and cartilage? How is it closed air-tight?

How is the chest closed above? How below? In what diameters can the chest cavity be enlarged?

Describe the diaphragm.

the dome, striped muscular fibres radiate, downwards and
outwards, in all directions, and are fixed by their inferior
ends to the lower ribs, the breast-bone, and the vertebral
column. In inspiration (drawing a breath) the muscular

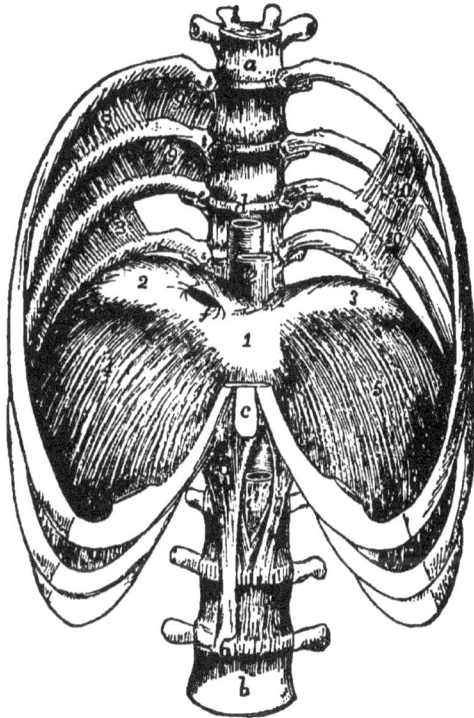

FIG. 69.—The lower half of the thorax, with four lumbar vertebræ, showing
the diaphragm from above. 1, 2, 3, central tendon; 4, 5, muscular part.

fibres, shortening, flatten the dome and so enlarge the
thoracic cavity at the expense of the abdominal. The contraction of the diaphragm thus greatly increases the size of
the thorax chamber by adding to its lowest and widest
part.

What happens to it during inspiration? What results?

The Dorso-Ventral Enlargement of the Thorax.—The ribs on the whole slope downwards (Fig. 15) from the vertebral column to the breast-bone, the slope being most marked in

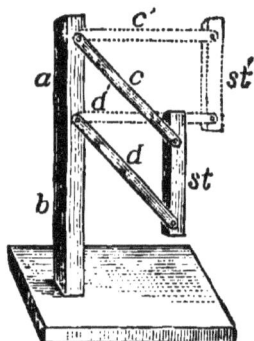

the lower ones. During inspiration the breast-bone and the sternal ends of the ribs attached to it are raised by muscles which pull on them, and so the distance between the sternum and the vertebral column is increased. That this must be so will readily be seen by examining the diagram Fig 70, where *ab* represents the vertebral column, *c* and *d* two ribs, and *st* the sternum. The continuous lines represent the natural position of the ribs at rest in expiration, and the dotted lines the position in inspiration. It is clear that when their lower ends are raised so as to make the bars lie in a more horizontal plane, the sternum is pushed away from the spine, and so the extent of the chest cavity between back-bone and breast-bone is increased.

FIG. 70.—Diagram illustrating the dorso ventral increase in the diameter of the thorax when the ribs are raised.

Expiration.—Inspiration requires a good deal of muscular effort. When the diaphragm contracts and flattens its dome it has to push down the abdominal viscera on its under side, and to press out the front wall of the abdomen to make room for them. The ribs and breast-bone have also to be pulled up out of their natural position of equilib-

What is the general direction of the ribs? How is it altered during inspiration? What is the consequence? Illustrate by reference to a diagram.

Why does inspiration require muscular effort?

rium. In ordinary expiration, on the contrary, but little if any muscular effort is required.

As soon as the muscles which have raised the ribs and sternum relax, these bones return to their natural uncon-strained position; and the elastic abdominal wall presses the abdominal viscera against the under side of the dia-phragm and pushes that organ up again, as soon as its muscular fibres cease contracting. In this way the chest cavity is restored to its original capacity, and the air is sent out of the lungs rather by the elasticity of the parts which were stretched in inspiration, than by any special expiratory muscles.

When, however, an expiration is violent (when, for ex-ample, we try to empty our lungs of air as completely as possible, or during a fit of coughing) special expiratory muscles, which pull down the ribs and press up the dia-phragm, are called into action.

The Respiratory Sounds —The entry and exit of air are accompanied by the *respiratory sounds* or *murmurs*, which can be heard on applying the ear to the chest wall. The character of these sounds is different and characteristic over the trachea, the larger bronchial tubes, and portions of lung from which large bronchial tubes are absent. They are variously modified in pulmonary affections, and hence the value of *auscultation* of the lungs in assisting the phy-sician to form a diagnosis.

Hygienic Remarks.—Since the diaphragm when it con-

How does expiration differ from it in this respect?
Explain how the chest is brought back to its resting position after an inspiration?
Give examples of violent expiration? How does it differ from an ordinary expiration in the forces at work for its production?
What are the respiratory sounds? Where do their characters differ? Why do physicians study them in lung diseases?

tracts pushes down the abdominal viscera lying against its under side, these have to make room for themselves by pushing out the soft front of the abdomen, which accordingly protrudes when the diaphragm descends. Hence breathing by the diaphragm is indicated on the exterior by

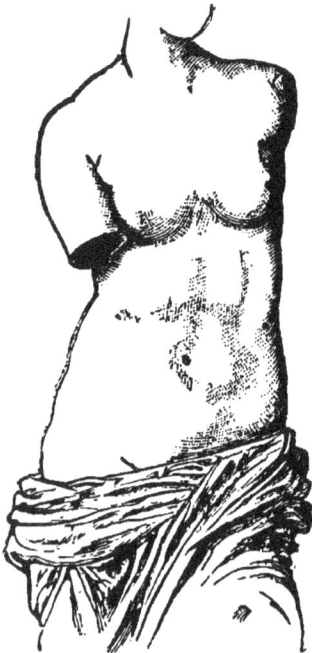

FIG. 71.

FIG. 71.—Torso of the Statue known as Venus of Milo.

FIG. 72.

FIG. 72.—Paris Fashion, May, 1880.

movements of the abdomen, and is often called "abdominal respiration," as distinguished from breathing by the ribs, called "costal" or "chest breathing." In both sexes the diaphragmatic breathing is the more important, but, as a rule, men and children use the ribs less than adult women.

Why does the abdomen protrude during an inspiration? What is meant by abdominal respiration? Why? What is costal respiration? Which form of breathing is most developed in women?

Since abdomen and chest alternately expand and contract in healthy breathing, anything which impedes their free movement is to be avoided: the tight lacing which used to be thought elegant, and, indeed, is still indulged in by some who think a distorted form beautiful, seriously impedes one of the most important functions of the body, leading not only to shortness of breath and an incapacity for muscular exertion, but, as has been proved, in many cases to actual disease. In extreme cases of tight lacing some organs are often directly injured, weals of fibrous tissue being, for example, not unfrequently found developed on the liver from the constant pressure of the lower ribs forced against it by a tight corset.

The Lungs, from the congested state of their vessels produced by alcohol, are more subject to the influence of cold, the result being frequent attacks of *bronchitis.* It has also been recognized of late years that there is a peculiar form of consumption of the lungs which is very rapidly fatal, and found only in alcohol-drinkers.

Why should conditions impeding the movement of chest and abdomen be avoided? What is the result of tight lacing? What is often found on examining after death the bodies of persons who have practised tight lacing for a long time?

APPENDIX TO CHAPTER XVII.

1. A sheep's lungs with the windpipe attached may be readily obtained from a butcher. It is best to secure it and the heart all in one mass, as unless the heart be carefully removed holes are apt to be cut in the lung

2. Examine the windpipe, and trace it down to its division into the bronchi. In the wall of the windpipe note the horse shoe shaped cartilages which keep it open, and which are so arranged that the dorsal aspect of the tube (which lies against the gullet) has no hard parts in it.

3. Trace one bronchus to its lung, and then cutting away the lung tissues follow the branching bronchial tubes through the organ. Note the cartilages in their walls.

4. Carefully divide the other bronchus where it joins the windpipe, and lay it and its lung aside. Then slit open the trachea, the bronchus still attached to it and the bronchial tubes. Observe the soft pale-red mucous membrane lining them.

5. In the bronchus which has still an uninjured lung attached to it tie air-tight a few inches of glass or other tubing of convenient size. On the end of the glass tube then slip a few inches of rubber tubing. On blowing through the rubber tube the lung will be distended, and as soon as the opening is left free it will collapse; in this way its great extensibility and elasticity will be seen.

6. Blow up the lung moderately, and while it is distended tie a string very tightly around the bit of rubber tubing. This will keep the air from escaping; the distended lung can now be examined at leisure, and its form, lobes, and the smooth moist pleura covering it be better seen than when it is collapsed.

7. To construct the very instructive model depicted in Fig. 67, obtain a wide necked glass vessel, and a rubber toy balloon. Very carefully untie and open the neck of the balloon, and tie into it tightly a glass rod. Take a cork (one of rubber is best) which fits the neck of the bottle tightly and is perforated by two holes; through one of these holes pass the tube projecting from the neck of the balloon in such way that the collapsed balloon is on the under side of the cork. Through the other hole pass, air tight, a tube bent as shown in the figure. and on the upper end of this slip a few inches of rubber tubing. (This can be pinched or tied up at any time, and in that way closed, and so forms a cheap substitute for the stopcock represented in the figure). When the cork is now secured firmly in the bottle the apparatus is ready for use as indicated on p. 269.

8. Substitute a lung for the rubber balloon in the above experiment.

9. The action of the diaphragm may be illustrated by substituting for the bottle of § 7 a bell-jar with a wide neck at its upper end. Take a piece of sheet rubber somewhat larger than the bottom of the bell-jar, and tie a button or marble in the centre of it. Lay the rubber on the table, with the projection caused by the button downwards. On it place the bell jar, stretch the rubber moderately tight, turn its edges up around the margin of the bell-jar, and tie very tightly with waxed cord or copper wire. In the neck of the bell-jar place a tight cork with tubes and rubber balloon, as described in § 7. Suck air out of the bottle until the balloon is fairly well expanded; then tie the rubber tube. As the air is removed the pressure

of the atmosphere on its exterior will cause the rubber sheet to arch up into the cavity of the bell-jar so that it now fairly well represents the diaphragm. The knob caused by the button serves as a handle by which this artificial diaphragm may be pulled down, representing inspiration; as it descends the balloon (lung) enlarges, and air enters it from outside. When the button is let go the artificial diaphragm ascends, the lung collapses, and air is forced out of it (expiration). Then open the air tube leading into the bell-jar. The lung will collapse, and the movements of the diaphragm have no influence upon it.

10. The diaphragm itself may be readily seen on the body of any small animal (rat, kitten, puppy), on removing the abdominal viscera. The liver and stomach must be cut away with especial care.

a. When the above viscera are removed the vaulted diaphragm will be seen, and through it the pink lungs.

b. Seizing some of the folds of peritoneum attached to the diaphragm, pull it down, imitating its contraction and flattening in inspiration. The lungs will be seen to follow it closely, expanding to fill the space left by it in its descent.

c. Make a free opening into one side of the thorax. The corresponding lung will collapse, and be no longer influenced by movements of the diaphragm

d. Now open the other side of the chest: its lung also shrinks up; the structure of the diaphragm (its tendinous centre and muscular peripheral regions) can now be better seen, as also the attachment of the pericardium to its thoracic side.

11. The action of the microscopic cilia in driving along the mucus in which they move may be demonstrated as follows:

a. Cut off a frog's head and destroy its spinal cord (p. 258). Then cut out its gullet as completely as possible; slit this open and spread it out, inner side up, on a piece of cork or board, and fix it with pins stuck through its edges.

b. Prepare a very thin and small shaving of cork. Dissect the skin off one thigh of the animal and wrap a bit of it round the shaving of cork, with its under side outwards.

c. Place the light mass thus formed on the mucous membrane of the gullet, near its mouth end. The little mass will slowly be moved along to the stomach end of the gullet, and if returned to the mouth end time after time will be swept along in the same direction. This is due to the cilia which line the frog's gullet (they are not present in that of man), and push along to the stomach the mucus bathing them on which the little float swims.

d. Place the exposed gullet under a bell-jar with a wet sponge for an hour or two. The mucus secreted by it will be found to have been swept along to the end of it which joins the stomach.

CHAPTER XVIII.

THE CHEMISTRY OF RESPIRATION AND VENTILATION.

The Quantity of Air breathed daily.—After an ordinary expiration the chest cavity is by no means completely collapsed. At this time the lungs still contain about 200 cubic inches of air. In the next inspiration 30 more cubic inches are taken in, about the same amount sent out at the following expiration, and so on throughout the day. During quiet breathing the quantity of air in the lungs varies, therefore, with each inspiration and expiration between 230 and 200 cubic inches. At each inspiration something over a pint of fresh air is taken in, and at each expiration about the same amount of vitiated air is expelled. As each of us breathes at least fifteen times a minute, we thus use each minute, and render impure, $15 \times 30 = 450$ cubic inches ($15\frac{1}{2}$ pints) of air. In an hour the quantity would be $450 \times 60 = 27,000$ cubic inches (930 pints), and in twenty-four hours $27,000 \times 24 = 648,000$ cubic inches (22,320 pints) of air, which would weigh about 28 7 lbs. We have next to

How much air do the lungs contain after an ordinary expiration? How much do they receive at the next inspiration? How much is sent out from them during expiration? Within what limits does the quantity of air in the lungs vary during quiet breathing? What bulk of air is taken in during an inspiration?

How often do we breathe? How much air does each one of us render impure every minute? How much in an hour? How many pints in a day? What does this quantity of air weigh?

see what it is that happens to this vast quantity of air breathed daily by each one of us; what we have taken out of it, and what we have given off to it.

The Changes produced in Air by being once breathed.— These are fourfold—changes in its temperature, in its mois‑ ture, in its chemical composition, and in its volume.

Temperature Changes.—The air taken into the lungs is nearly always cooler than that expired, which has a tem‑ perature of about 36° C. (97° F.). The temperature of a room is usually about 21° C. (70° F.). The warmer the inspired air, the less the heat which is lost to the body in the breathing process.

Changes in Moisture.—Inspired air always contains more or less water vapor, but is rarely saturated—that is, rarely contains so much but it can take up more without showing it as mist; the warmer air is, the more water vapor it re‑ quires to saturate it. The expired air is nearly saturated for the temperature at which it leaves the body, as is read‑ ily shown by the vapor deposited when it is slightly cooled, as when a mirror is breathed upon; or by the clouds seen issuing from the nostrils on a frosty day, these being due to the fact that the air as soon as it is cooled cannot hold all the water vapor which it took up when warmed in the body. We therefore conclude that air when breathed gains water vapor and carries it off from the lungs. The quan‑ tity of water thus removed from the body is about nine ounces each twenty-four hours.

Chemical Changes.—The most important changes brought about in the breathed air are those in its chemical

What changes are produced in the air on its being once breathed? How does expired air differ in temperature from inspired? How does expired air differ from inspired in moisture? How much water is evaporated from the lungs daily?

composition. Pure air when completely dried consists in 100 parts of—

	By Volume.	By Weight.
Oxygen	20.8	23
Nitrogen	79.2	77

When breathed once, such air gains rather more than 4 volumes in 100 of carbon dioxide, and loses rather more than 5 of oxygen. More accurately, 100 volumes of expired air, when dried, consist of—

Oxygen	15.4
Nitrogen	7.2
Carbon dioxide	4.3

The expired air also contains volatile organic substances in quantities too minute for chemical analysis, but readily detected by the nose upon coming into a close room in which a number of persons have been collected.

The Quantity of Oxygen taken up by the Lungs in a Day.—We have already seen that the quantity of air breathed in a day is 648,000 cubic inches. This loses 5.4 per cent. of oxygen or 35,000 cubic inches, weighing 12,818 grains (1⅚ lbs.): the body therefore gains this amount of that gas through the lungs daily.

The Amount of Carbon Dioxide passed out from the Lungs in a Day.—This being 4.3 per cent. of the total bulk of the air breathed, is 27,864 cubic inches; it weighs 14,105 grains or about 2 lbs.

We thus find that though each breath seems in itself a

What is the chemical composition of pure air by volume? By weight? What substances does air gain and lose when once breathed? What is the composition by volume of dried expired air? What does dried expired air contain besides oxygen, nitrogen, and carbon dioxide?

What bulk of oxygen does the air breathed in a day lose? What weight? How much oxygen does the body take up daily by means of the lungs?

What bulk of carbon dioxide is carried off by breathing in each day? What does it weigh?

very little thing, on calculation it is obvious that the total amount of matter received into the body from the lungs, and that passed out of it by these organs, every day of our lives is considerable. In a year each adult breathes about 10,000 lbs. of air ; from it he takes 657 lbs. of oxygen, and to it he gives off 730 lbs. of carbon dioxide.

Changes of Volume in Air once breathed.—If the expired air be measured as it leaves the body its bulk will be found greater than that of the inspired air, since it not only has water vapor added to it, but is expanded in consequence of its higher temperature. If, however, it be dried and reduced to the same temperature as the inspired air, its volume will be found diminished, since it has lost 5.4 volumes of oxygen for every 4 3 volumes of carbon dioxide which it has gained.

Ventilation.—Since at each breath some oxygen is taken from the air and some carbon dioxide given to it, were the atmosphere around a living person not renewed he would at last be unable to get from the air the oxygen he required; he would die of oxygen starvation or be *suffocated,* as such a mode of death is called, as surely, though not quite so fast, as if he were put under the receiver of an air-pump and all the air around him removed. Hence the necessity of *ventilation* to supply fresh air in place of that breathed, and clearly the amount of fresh air requisite must be determined by the number of persons collected in a room: the supply which would be ample for one person would be in-

What weight of air is breathed yearly by an adult? How much oxygen is taken from it? How much carbon dioxide is given to it?
How does the air expired differ in bulk from inspired? Why? If the expired air be dried and cooled to the temperature of the inspired, what is found? Why?
Why would a man die if the air around him were not renewed? What is suffocation? What is the object of ventilation?

sufficient for two. Moreover, fires, gas, and lamps all use up the oxygen of the air and give carbon dioxide to it, and hence calculation must be made for them in arranging for the ventilation of a building in which they are to be used.

When breathed Air becomes unwholesome.—In order that air be unwholesome to breathe, it is by no means necessary that it shall have lost so much of its oxygen as to make it difficult for the body to get what it wants of that gas. The evil results of insufficient air-supply are rarely directly due to that cause even in the worst ventilated room, for the blood flowing through the lungs can take what oxygen it wants from air containing comparatively little of that gas. The headache and drowsiness which come on from sitting in badly ventilated rooms, and the want of energy and general ill-health which result from permanently living in such, are dependent on a slow poisoning of the body by the reabsorption of matters eliminated from the lungs in previous respirations. What these are is not accurately known; they doubtless belong to those volatile bodies mentioned above as carried off in small quantities in each breath, since observation shows that the air becomes injurious long before the amount of carbon dioxide in it is sufficient of itself to do any harm. Breathing air containing one or two per cent. of that gas produced by ordinary chemical methods does no particular injury, but the breathing of air containing one per cent of carbon

What conditions determine the supply of fresh air which should be provided to a room?
Is air ever unwholesome while still capable of supplying the oxygen which the body requires? What results from living in ill-ventilated rooms? Why? Does air once breathed become injurious before the quantity of carbon dioxide in it is poisonous? What percentage of pure carbon dioxide may be present in the air breathed without doing harm?

dioxide produced by respiration is decidedly injurious, because of the other things sent out of the lungs along with it. Carbon dioxide, in any such percentage as is commonly found in a room, is not poisonous, as used to be believed, but as it is tolerably easily estimated in air, while the more dangerous injurious substances evolved in breathing are not, the purity or foulness of the air in a room is usually determined by finding the percentage of carbon dioxide in it; but it must be borne in mind that to mean much this carbon dioxide must have been produced by breathing; otherwise the amount of it present is no guide to the quantity of the more important poisonous substances present. Of course when a great deal of carbon dioxide is present the air is irrespirable, as for example sometimes at the bottom of wells or brewing-vats.

The Quantity of Fresh Air which should be allowed for each Person in a Room.—In each minute a man breathes out 450 cubic inches of air containing rather more than 4 per cent. of carbon dioxide. This mixed with three times its bulk of pure air would give a little over one cubic foot containing one per cent. of carbon dioxide. Such air is no longer respirable with safety. The result of breathing it for an hour or two is headache and drowsiness; of breathing it for weeks or months several hours daily, a

What percentage of carbon dioxide produced by breathing shows that the air is unfit for use? Is the proportion of carbon dioxide found ordinarily in a room poisonous? Why is the percentage of carbon dioxide in air usually employed in deciding whether the air is fit to breathe? What must be borne in mind in deciding from the percentage of carbon dioxide in it that air is no longer wholesome? Is air containing much carbon dioxide fit to breathe? Give an example.

What bulk of air does a man contaminate with carbon dioxide to the extent of one per cent. in each minute? Is air so contaminated fit to breathe? What are the consequences of breathing it for a few hours? What of breathing it for months?

lowered tone of the whole body, less power of work, physical or mental, and less power of resisting disease. The ill effects may not show themselves at once, and may accordingly be overlooked or considered scientific whims by the careless, but they are there ready to manifest themselves nevertheless.

In order to have air to breathe in a fairly pure state every man should have for his own allowance at least about 800 cubic feet of space to begin with, and the arrangements for ventilation should at the very least renew this at the rate of one cubic foot per minute. The nose is, however, the best guide, and it is found that at least five times this supply of fresh air is necessary to keep free from any odor the room inhabited by one adult. If an inhabited room smells " close" to one coming into it from " out of doors," the air in it is unwholesome to breathe for any length of time.

How to Ventilate.—Ventilation does not necessarily mean draughts of cold air, as is too often supposed. In warming by indirect radiation it may readily be secured by fixing, in addition to the registers from which the fresh warmed air reaches the room, corresponding openings at the opposite side by which the old air may pass off to make room for the new. An open fire in a room will always keep up a current of air through it, and is one of the most wholesome, though not most economical, methods of warming an apartment.

Why are the injurious effects of impure air apt to be ignored?
What volume of air should be allowed to each adult? At what rate should it be replenished? What supply of fresh air is needed to keep an inhabited room free from odor? Is it safe to live in a room which " smells close"?
How may ventilation be secured in heating by "indirect radiation"? What are the advantages and what the disadvantages of an open fire?

Stoves in a room unless constantly supplied with fresh air from without dry its air to an unwholesome extent. If no appliance for providing this supply exists in a room it can usually be got, without a draught, by fixing a board about four inches wide under the lower sash and shutting the window down on it. Fresh air then comes in by the opening between the two sashes and in a current directed upwards, which gradually diffuses itself over the room without being felt as a draught at any one point. In the method of heating by direct radiation the apparatus employed provides of itself no means of drawing fresh air into a room, as the draught up the chimney of an open fireplace or of a stove does; and therefore special inlet and outlet openings are very necessary. Since, fortunately, few doors and windows fit quite tight, fresh air gets into closed rooms in tolerable abundance for one or two inhabitants.

Changes undergone by the Blood in the Lungs.—These are the exact reverse of those exhibited by the breathed air —what the air gains the blood loses, and *vice versa.* The blood loses heat, and water, and carbon dioxide in the pulmonary capillaries, and gains oxygen. These gains and losses are accompanied by a change of color from the dark purple which the blood exhibits in the pulmonary artery, to the bright scarlet it possesses in the pulmonary veins.

Why the Blood changes its Color as it flows through

What are stoves apt to do? Point out a good way of supplying fresh air to a room warmed by a stove? What are especially important in a room heated by "direct radiation"? Why is it fortunate that doors and windows do not fit air-tight?

What is the relationship between the gains and losses of blood in the pulmonary circulation, and the losses and gains of the breathed air? What does the blood lose as it flows through the lungs? What does it gain? What change in the color of the blood accompanies these gains and losses?

the **Pulmonary Capillaries.**—The color of the blood depends on its red corpuscles, since pure blood-plasma or blood-serum is colorless or at most a very faint straw yellow. Hence the color change which the blood experiences in circulating through the lungs must be due to some change in its red corpuscles. These consist chiefly of *hæmoglobin* (p. 204), and hæmoglobin, as we have learned, is a substance which has the power of absorbing oxygen and forming a bright scarlet compound called *oxyhæmoglobin*. This oxyhæmoglobin very easily gives up its oxygen when it is placed under conditions where that gas is scarce: the hæmoglobin left behind has a dark purple color. The blood leaving the lungs by the pulmonary veins is bright red because all its hæmoglobin has been turned into oxyhæmoglobin. From the left side of the heart it is conveyed by the branches of the aorta to all the organs of the body. These are constantly using oxygen, which is therefore very scarce in them, and as the blood flows through its oxyhæmoglobin is broken up, the oxygen taken away, and dark purple-red hæmoglobin left to be conveyed by the veins to the right auricle of the heart. From there it passes to the right ventricle and thence by the pulmonary artery to the lungs, where it again picks up oxygen and

Why must the change in color of the blood during its pulmonary circuit be due to a change in its red corpuscles? What is the chief constituent of the red blood-corpuscles? What property does hæmoglobin possess with reference to oxygen? What is the color of oxyhæmoglobin?

When does oxyhæmoglobin give off its oxygen? What is the color of hæmoglobin? Why is the blood in the pulmonary veins bright red? Where is it conveyed? Why is oxygen scanty in most organs of the body? What results as regards the oxyhæmoglobin of the blood? What is the color of the blood sent to the right auricle of the heart? What is the subsequent course of this blood until it reaches the lungs? What does it receive in the lungs? How is its color altered in those organs?

becomes bright-red oxyhæmoglobin. The red corpuscles of the blood are so many little boxes in which oxygen is packed away in the lungs for conveyance to distant parts of the body.

What is the function of the red blood-corpuscles?

APPENDIX TO CHAPTER XVIII.

1. To show that air is warmed by breathing, breathe for a few seconds on the bulb of a thermometer. The mercury will be seen to rise rapidly in the stem.

2. To demonstrate that air gains water in the lungs, breathe on a mirror, or on a knife-blade or other polished metallic surface.

3. The presence of carbon dioxide in expired air may be readily demonstrated by expiring through a tube immersed in lime-water. This may be obtained at any drug-store; with carbon dioxide it gives a white precipitate, which dissolves readily in a little vinegar.

4. To show that much less carbon dioxide exists in inspired air, take a small bottle with a wide neck. Fit tightly into the neck of the bottle a cork perforated by two holes. Through one hole pass a glass tube reaching to near the bottom of the bottle, and through the other one which ends just below the cork; on the outer end of this tube fit a foot or so of rubber tubing. Remove the cork; half fill the bottle with lime-water and then replace the cork. Suck air through the rubber tubing. It will bubble through the lime-water, but (unless the room is very badly ventilated) a great deal must be drawn through the lime-water before as abundant a precipitate is produced as that which results from blowing a small quantity of breathed air (3) through the lime-water.

5. The influence of oxygen upon the color of the blood may be illustrated as follows:

a. Take to a slaughter-house a glass jar or beaker (an ordinary tumbler answers quite well), two bottles, an earthenware quart pitcher, and a bundle of wire.

b. When an animal is killed and bled, collect some blood in the jar and let it clot.

c. Collect some more blood in the pitcher, and defibrinate by thorough whipping with the bundle of wire. Half fill each bottle with the defibrinated blood; then cork the bottles.

d. Having brought home all the specimens, set them aside until the next morning in a cool place. It will then be found that the blood in each bottle is dark-colored or venous (having used up its own oxygen and not being able to get more from the air), and that the clot in the jar is bright scarlet (arterial-colored) above where it is in contact with the air, but dark purple-red where it is immersed in the serum.

e. Invert the clot: in an hour or two its previously dark original under surface will have become bright red, while the original upper surface, previously bright-colored and now immersed in the serum away from the air, will have become venous in tint.

f. Take the cork out of one bottle; renew the air in it by blowing. Placing a thumb on the neck of the bottle, thoroughly shake up the blood with the air. Then renew the air again, and shake once more; and so on for three or four times. At the end the blood shaken up with air will be seen to have assumed a much brighter red color than that kept shut up in the other bottle.

g. If the proper chemical apparatus and reagents are accessible, the air in the bottle about to be shaken may be replaced by nitrogen, hydrogen, or pure oxygen, and the procedures described in section *f* repeated. It will be found that only the oxygen brightens the blood color. As any one possessing the chemical apparatus and knowledge implied for the execution of this experiment will certainly know how to replace the air by the gases above named, no further details need be given.

CHAPTER XIX.

THE KIDNEYS AND THE SKIN.

General Arrangement of the Nitrogen-excreting Organs.—These organs are (1) the *kidneys*, the glands which secrete the urine; (2) the *ureters* or ducts of the kidneys, which carry the secretion to (3) the *urinary bladder*, a reservoir in which it accumulates and from which it is expelled from time to time through (4) an exit tube, the *urethra*. The general arrangement of these parts, as seen from behind, is shown in Fig. 73. The kidneys, *R*, lie at the back of the abdominal cavity, opposite the upper lumbar vertebrae, one on each side of the middle line. Each is a solid mass, with a convex outer and a concave inner border, and its upper end a little larger than the lower. From the abdominal aorta, *A*, a *renal artery*, *Ar*, enters the inner border of each kidney, to break up within it into finer branches, ultimately ending in capillaries. The blood is collected from these into the *renal veins*, *Vr*, one of which leaves each kidney and opens into the inferior vena cava, *Vc*, which carries it, after having lost water and urea in the kidney, back to the heart. From the concave

Name the chief organs concerned in removing from the body its nitrogenous waste matters. Describe the general arrangement of these organs. Describe the form of a kidney. What vessel supplies it with blood? What vessel carries blood out of the kidney? What has the blood carried off from a kidney lost while flowing through that organ?

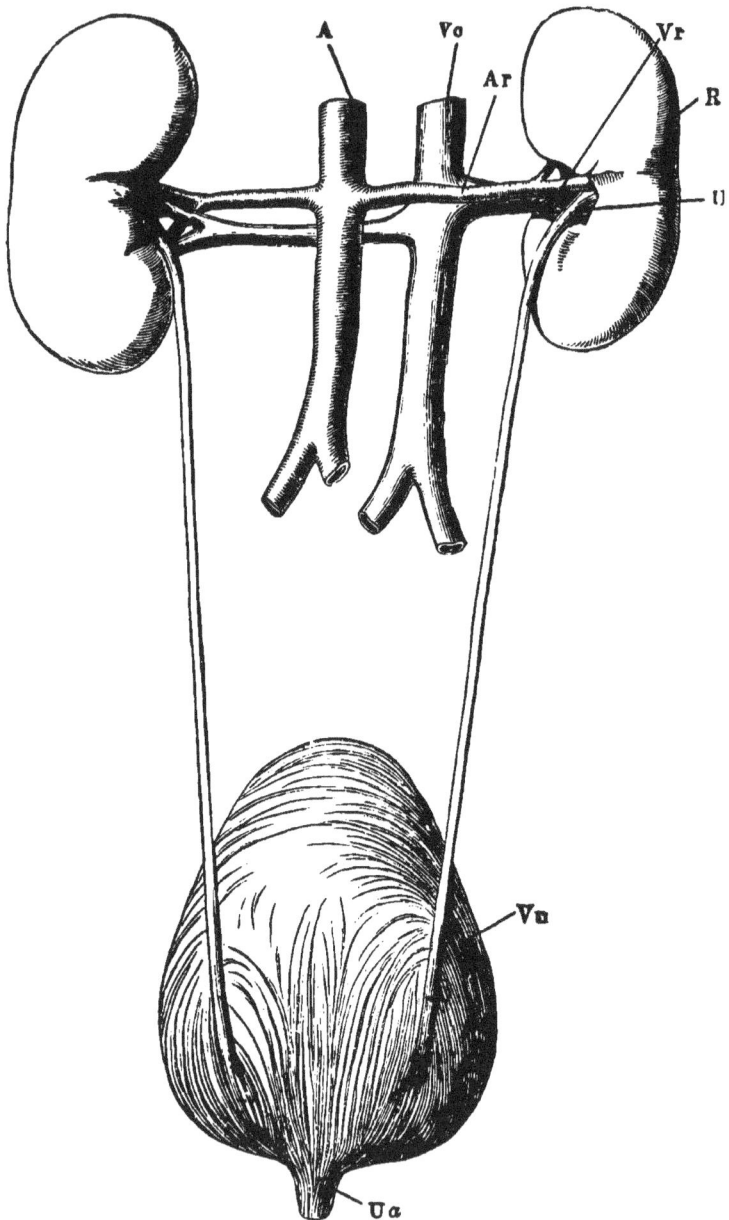

Fɪɢ. 73.—The renal organs, viewed from behind. R, right kidney; A, aorta; Ar, right renal artery; Vc, inferior vena cava; Vr, right renal vein; U, right ureter; Vu, bladder; Ua, commencement of urethra.

border of each kidney proceeds also the *ureter*, *U*, a slender tube opening below into the bladder, *Vu*, near its lower end. From the bladder proceeds the urethra, at *Ua*. The channel of each ureter passes very obliquely through the wall of the bladder; accordingly if the pressure inside the latter organ rises above that of the liquid in the ureter the walls of the oblique passage are pressed together and it is closed. Usually the bladder (which contains muscular tissue in its walls) is relaxed, and the urine flows readily into it from the ureters. The commencement of the urethra being kept closed by elastic tissue around it (which can voluntarily be reinforced by muscles which compress the tube), the urine accumulates in the bladder. When this latter contracts and presses on its contents the ureters are closed in the way above indicated, the elasticity of the fibres closing the urethral exit from the bladder is overcome, and the liquid is forced out.

Naked Eye Structure of the Kidneys.—When a section is made through a kidney from its outer to its inner border (Fig. 74) it is seen that a deep fissure, the *hilus*, leads into the latter. In the *hilus* the ureter widens out to form the *pelvis of the kidney*, which breaks up into a number of smaller divisions, the cups or *calices*. The cut surface of the kidney proper is seen to consist of two distinct parts; an outer or *cortical portion*, and an inner or *medullary*. The medullary portion is less red and more glistening to the eye, is finely striated in a radial direction, and does not

What is the ureter? Under what circumstances is its opening into the bladder closed? What is the usual state of things? How is the commencement of the urethra closed? What results? What happens when the bladder contracts?

What is seen on a section made through a kidney? How is the pelvis of the kidney formed? What are the calices? What is seen on the cut surface of the kidney proper? How does the medullary part differ in appearance from the cortical?

consist of one continuous mass, but of a number of conical portions, the *pyramids of Malpighi,* 2′, each of which is

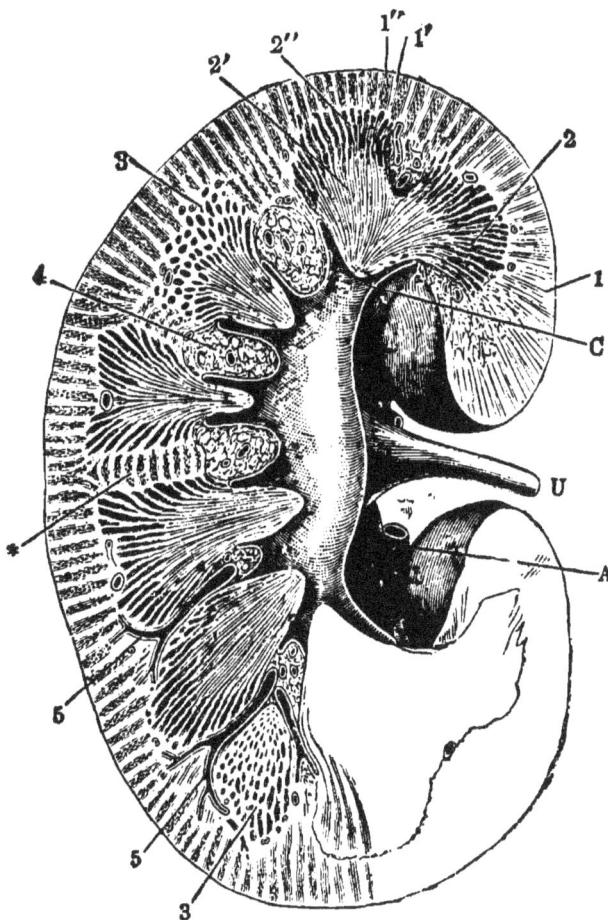

FIG. 74.—Section through the right kidney from its outer to its inner border. 1, cortex; 2, medulla; 2′, pyramid of Malpighi; 2″, pyramid of Ferrein; 5, small branches of the renal artery entering between the pyramids; *A,* a branch of the renal artery; *C,* the pelvis of the kidney; *U,* ureter; *C,* a calyx.

separated from its neighbors by a prolongation,*, of the cortical substance. This, however, does not reach to the

Describe the pyramids of Malpighi. What lies between them?

apex of the pyramid, which projects, as the *papilla,* into a calyx of the ureter. At its outer end each pyramid separates into smaller portions, 2″, separated by thin layers of cortex and gradually spreading everywhere into the latter. The cortical substance is redder, more granular looking, and less shiny than the medullary; it forms everywhere the outer layer of the organ, besides dipping in between the pyramids in the manner above described.

The renal artery divides in the hilus into branches (5) which run into the kidney substance between the pyramids, give off a few twigs to the pyramids, and end finally in a much closer vascular network in the cortex.

The Minute Structure of the Kidney.—The kidneys are compound tubular glands, being composed of branched microscopic *uriniferous tubules,* lined by a single layer of secreting cells, supported by connective tissue, and supplied with blood-vessels, nerves, and lymphatics. The final branches of each tubule end in a dilatation which contains a knot of blood-vessels, through whose walls water and salts filter into the tubule. As the water trickles along the latter, the cells lining it pass out the nitrogenous wastes of the blood brought there by the capillaries which wrap closely around them. The tubules unite in the pyramids to form fewer and larger ducts, which pour the secretion

How does the apex of a pyramid end? Where do we find the cortical substance of the kidney?

Describe the general distribution of the renal artery and its branches in the kidney. What part of the kidney contains most capillary blood-vessels?

To what type of gland do the kidneys belong? Of what are they made up? How do the uriniferous tubules end? What lies in each dilatation? What filters from the blood-vessels of its dilatation into the cavity of the tubule? What is added to this as it trickles along the tubule? How is the nitrogenous waste of the body brought close to the kidney tubules? What becomes of the tubules in the pyramids? Where do the larger ducts convey the secretion?

into the calices of the pelvis of the ureter, and this tube then conveys it to the bladder.

The Renal Secretion is less in bulk in warm weather, when perspiration carries off a good deal of the excess water of the blood, than in cold. On an average the kidneys eliminate from the body in twenty-four hours about fifty ounces (2½ pints) of water, and 500 grains (1⅐ ounces) of urea, which contain a little more than 230 grains of nitrogen.

The Kidneys, being the chief organs for the excretion of nitrogen waste, are among the most important organs of the body. Any defect in their healthy activity leads to serious interference with the working of many organs, due to the accumulation in the body of nitrogenous waste products.*

Hygiene of the Kidneys.—If both kidneys be cut out of an animal, it dies in a few hours from blood-poisoning, due to accumulation of waste poisonous substances which the kidneys should have got rid of. Serious kidney-disease amounts to pretty much the same thing as cutting out the organs, since they are of little use if not healthy. It is always fatal if not checked, and often kills in a short time. The things which most frequently cause kidney-disease are undue exposure to cold, and indulgence in alcoholic drinks.

Where does the ureter convey it?
Is the bulk of the renal secretion greater in summer or in winter? Why? What is its average daily amount? How much urea does it contain? How much nitrogen is contained in this quantity of urea?
Why are the kidneys important organs? What follows when their physiological work is defective?

---

* *Bright's Disease,* one of the commonest and most dangerous of maladies, consists essentially in an alteration of the kidney structure, in consequence of which these organs cease to eliminate urea from the blood, and drain off pure albumen from it instead. The three most common causes of Bright's disease are (1) an acute illness, as scarlet fever, of which it is a frequent result, (2) sudden exposure to cold when warm (this often drives blood in excessive amount from the skin to internal organs and leads to kidney disease); and (3) the habitual drinking of alcoholic liquids.

Cold Causes Kidney-Disease partly by driving blood from the surface and congesting the kidneys, partly by throwing too much work on them. When the skin does not get rid of its proper share of the waste matters of the body, it is the kidneys which have to make up the deficiency.

Nearly all the infectious diseases which are accompanied by a rash on the skin, as measles and scarlet fever, also affect the kidneys. During these diseases the kidneys are more or less inflamed, and in the early stages of recovery they are still weak and easily injured. Under these circumstances, exposure to cold is very apt to cause incurable kidney-disease.

Alcohol Causes Kidney-Disease in Several Ways.—In the first place it unduly excites the activity of the organs. Next, by impeding oxidation it interferes with the proper preparation of nitrogen wastes: they are brought to the kidneys in an unfit state for removal, and injure those organs. Third, when more than a small quantity of alcohol is taken, some of it is passed out of the body unchanged, through the kidneys, and injures their substance.

The kidney-disease most commonly produced by alcohol is one kind of " Bright's disease," so called from the physician who first described it. The connective tissue of the organ grows in excess, and the true excreting kidney-substance dwindles away. At last the organ becomes quite unable to do its work, and death results.

What is the consequence of removing the kidneys? Of kidney-disease? How is serious disease of the kidneys most often produced?

How does cold injure the kidneys? When have they to do the work of the skin? What diseases especially affect them? State of the kidneys during recovery from these diseases? Precautions to be taken?

State one effect of alcohol on the kidneys. Another? A third? What kidney-disease is commonly produced by alcoholic excess? How are the kidneys altered by it? Results?

The Skin, which covers the whole exterior of the body, consists everywhere of two distinct layers: an outer, the *cuticle* or *epidermis ;* and a deeper, the *dermis, cutis vera,* or *corium.* A blister is due to the accumulation of liquid between these two layers. *Hairs* and *nails* are excessively developed parts of the epidermis.

The Epidermis, Fig. 75, consists of cells, arranged in many layers, and united by a small amount of cementing substance. The deepest layer, *d,* is composed of elongated or *columnar* cells, set on with their long axes perpendicular to the corium beneath. It is succeeded by several strata of roundish cells, *b,* which in the outer layers become more and more flattened in a plane parallel to the surface. The outermost epidermic stratum is composed of many layers of very flattened cells from which the nuclei (conspicuous in the deeper layers) have disappeared. These superficial cells are dead, and are constantly being shed from the surface of the body, while their place is taken by new cells, formed in the deeper layers, pushed up to the surface, and flattened in their progress. The change in the form of the cells as they travel outwards is accompanied by chemical changes ; they finally constitute a semi-transparent dry *horny stratum, a,* distinct from a deeper, more opaque, and softer layer, *b* and *d,* of the epidermis.

The rolls of material which are peeled off the skin on rubbing it with a rough towel after a warm bath, are dead outer scales of the horny stratum of the epidermis.

Of what two main parts does the skin consist ? What is a blister ? What are hairs and nails?

Describe the epidermis.

What portion of it is constantly being cast off from the body ? What is meant by the horny stratum of the epidermis ? Of what do the rolls of matter consist which are peeled off the skin on vigorous rubbing after a warm bath ?

Nerves penetrate the epidermis, but no blood-vessels enter it. In dark races the color is due to a pigment contained chiefly in the deeper-placed cells of the epidermis.

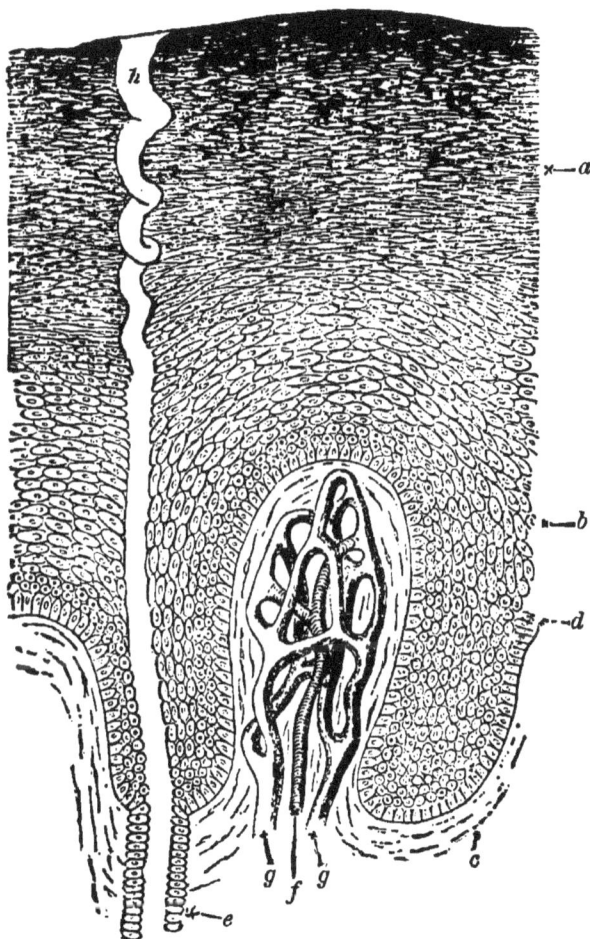

Fig. 75.—A section through the epidermis, somewhat diagrammatic, highly magnified. Below is seen a papilla of the dermis, with its artery, *f*, and veins, *gg*; *a*, the horny layer of the epidermis; *b*, the *rete muscosum* or Malpighian layer; *d*, the layer of columnar epidermic cells in immediate contact with the dermis; *h*, the duct of a sweat-gland.

Does the epidermis contain nerves? Blood-vessels?
Where does the coloring matter of the skin of negroes lie?

The Corium, Cutis Vera, or True Skin, Fig. 76, consists of a close feltwork of connective tissue, which, becoming wider meshed below, passes gradually into the *subcutaneous areolar tissue,* in texture much like damp raw cotton, which loosely attaches the skin to parts beneath. In tanning,

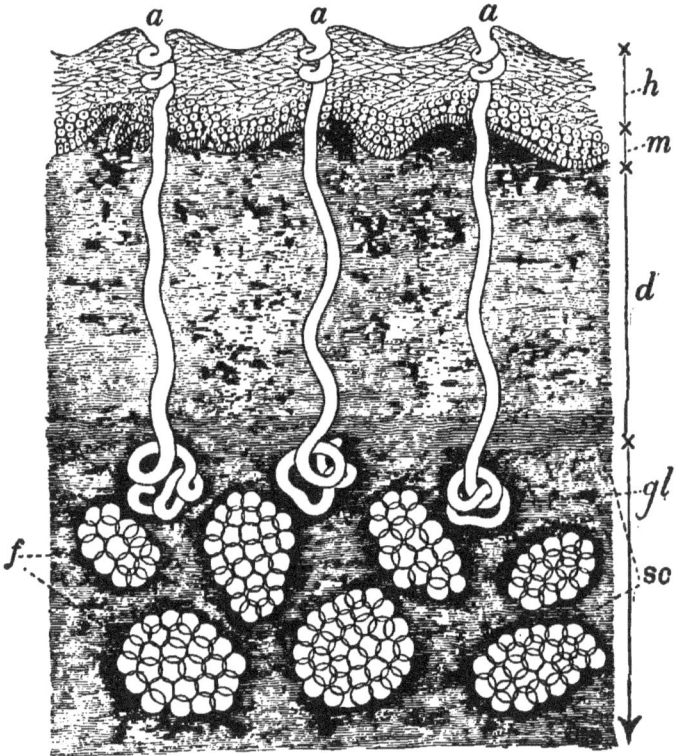

FIG. 76.—A section through the skin and subcutaneous areolar tissue. *h,* horny stratum, and *m,* deeper more opaque layer of the epidermis; *d,* dermis passing below into *sc,* loose areolar tissue, with fat, *f,* in its meshes: above, dermic papillæ are seen, projecting into the epidermis which is moulded on them. *a,* opening of a sweat-gland; *gl,* the gland itself.

the *cutis vera* is turned into leather, its connective tissue forming an insoluble and tough compound with the tannin of the oak-bark employed. Wherever there are hairs bundles of plain muscular tissue are found in the corium ; it contains also a close network of capillary blood-vessels, and

Of what does the true skin consist ? Into what does it pass on its deeper side ? What is turned into leather when an ox-hide is tanned ?

numerous lymphatics and nerves. In shaving, as long as the razor keeps in the epidermis there is no bleeding; but a deeper cut shows at once the presence of blood-vessels in the true skin.

The Papillæ of the Dermis.—The outer surface of the corium is almost everywhere raised into minute elevations, called *papillæ*, on which the epidermis is moulded, so that its deep side presents pits corresponding to the projections of the dermis. In Fig. 75 is shown a papilla of the *corium* containing a knot of blood-vessels, supplied by the small artery, *f*, and having the blood carried off from them by the two little veins, *gg*. Other papillæ contain no capillary loops, but, instead, special organs connected with nerve-fibres, and supposed to be concerned in the sense of touch. On the palm of the hand the dermic papillæ are especially well developed (as they are in most parts where the sense of touch is acute), and are frequently *compound* or branched at the tip. On this surface of the hand they are arranged in rows; the epidermis fills up the hollows between the papillæ of the same row, but dips down between adjacent rows, and thus are produced the finer epidermic ridges seen on the palms.* The wrinkles of old persons are due to the absorption of subcutaneous fat and of other soft parts beneath the skin, which, not shrinking itself to the same extent, is thrown into folds.

Hairs, longer or shorter, are found all over the surface of the body, except in a few regions, as the palms of the hands and the soles of the feet. A hair is a slender thread

How? Where do we find plain muscular tissue in the corium? What blood-vessels are found in it? What else?

What are the papillæ of the epidermis? Describe a papilla containing blood-vessels. What is found in other papillæ? Name a region of the skin where the dermic papillæ are especially developed. How are the ridges seen on the palm of the hand produced? To what are the wrinkles on the skin of elderly persons due?

---

* The more marked furrows of the palm, the so-called "lines of life" of the gypsy's palmistry, have a different origin.

of epidermis, developed on a special dermic papilla placed at the bottom of a depression, formed by a pitting-in of the dermis. The depression is known as the *hair follicle.* The part of the hair buried in the follicle is called its *root;* this is succeeded by a *stem*, which (in uncut hairs) tapers off to a *point.* Each hair is made up of a number of epidermic cells, arranged together so as to form a fibre.

Nails.—A nail is a part of the epidermis, with its horny stratum greatly developed. The back part of the nail fits into a furrow of the dermis, and is called its *root.* The visible part consists of a *body*, attached to the dermis beneath (which forms the *bed of the nail*), and of a *free edge.* Near the root is a little area, whiter than the rest of the nail, called the *lunula.* The whiteness is due in part to the nail being really more opaque there, and partly to the fact that its bed, which seen through the nail causes its pink color, is in this region less vascular.

The portion of the corium on which the nail is formed is called its *matrix.* Behind, this forms a groove lodging the root of the nail,

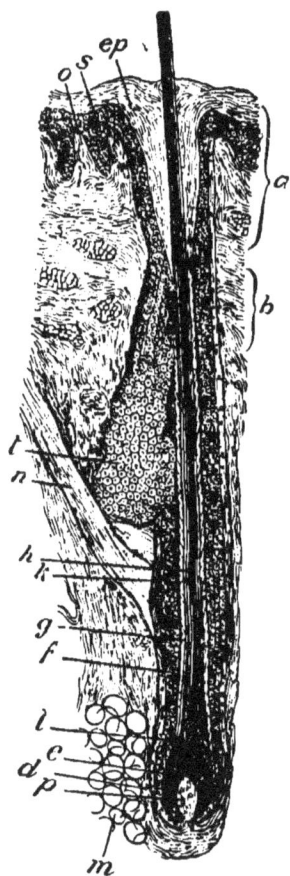

FIG. 77.—*ep*, horny layer of epidermis. lining mouth of follicle and continued down it; *s*, Malpighian layer; *o*, a papilla of the dermis; *l*, root of the hair; *p*, papilla of skin on which the hair grows; *t*, a sebaceous gland; *m*, fat cells.

Name regions of the skin which have no hairs. What is a hair? What is a hair follicle? What is the root of a hair? The stem? Of what is a hair composed?

What is a nail? Of what parts does it consist? What is the lunula? Why is it paler in color than the rest of the nail?

and it is by new cells added there that the nail grows in length. The part of the matrix lying beneath the body of the nail, and called its *bed*, is highly vascular; new cells formed on its bed and added to its under surface cause the nail to increase in thickness, as it is pushed forward by the new growth at its root. The free end of a nail is therefore its thickest part. If a nail is "cast" in consequence of an injury, or torn off, a new one is produced, provided the matrix is not destroyed.

The **Glands of the Skin** are of two kinds. the *sweat glands* or *sudoriparous glands,* and the *oil glands* or *sebaceous glands.*

The **Sweat Glands** (Fig. 78) are microscopic tubes which reach from the surface of the skin to the subcutaneous areolar tissue; then the tube often branches, and is coiled up into a little knot, intertwined with blood capillaries. These glands are found all over the skin, but are most abundant on the palms of the hands, the soles of the feet, and the brow. Altogether, there are about two and a half millions of them.

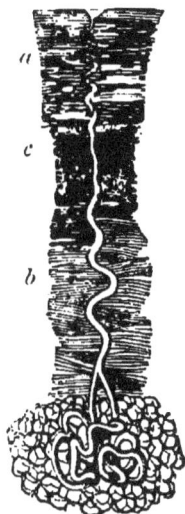

Fig. 78.—A sweat gland. *d*, horny layer of cuticle; *c*, Malpighian layer; *b*, dermis. The coils of the gland proper, imbedded in the subcutaneous fat, are seen below the dermis.

The **perspiration or sweat** poured out by the sudoriparous glands is a transparent colorless liquid, with a peculiar odor, varying in different races, and in the

What is the matrix of a nail? When does the nail grow longer? What is the bed of a nail? How does a nail grow thicker? Which is its thickest part? What is necessary in order that a "cast nail" may be reproduced?

What glands are found in the skin? Describe a sweat gland. Where are the sweat glands most numerous? How many are there on the whole skin?

same individual in different regions of the body. Its quantity in twenty-four hours is subject to great variations, but
usually lies between 10,850 and 31,000 grains (or 25 and 71
ounces). The amount is influenced mainly by the surrounding temperature, being greater when this is high ; but it is
also increased by other things tending to raise the temperature of the body, as muscular exercise.* The sweat may or
may not evaporate as fast as it is secreted ; in the former
case it is known as *insensible*, in the latter as *sensible perspiration*. By far the most passes off in the insensible
form, drops of sweat only accumulating when the secretion
is very profuse, or the surrounding atmosphere is so humid
that it does not readily take up more moisture. The perspiration in 1000 parts contains 990 of water to 10 of solids.
Among the latter is, in health, a little urea, some sodium
chloride, and other salts. In diseased conditions of the
kidneys the urea may be greatly increased, the skin sup-
plementing to a certain extent deficiencies of those organs.

　The Sebaceous Glands nearly always open into hair follicles. They are small compound racemose glands (p. 149).
Each presents a duct, opening near the mouth of the hair
follicle ; when followed back this duct is found to divide
into several branches which end in globular expansions

Describe the perspiration. How much is secreted daily ? Point
out conditions influencing its amount. What is meant by "insensible perspiration"? Under what conditions do we find "sensible
perspiration"? What percentage of solids exists in the sweat? What
do they contain ? When is the proportion of urea in the perspiration
apt to be increased ?
　Where do the sebaceous glands open ? To what type of gland do
they belong?

---

* In fever the sweat glands are paralyzed, and we find a high temperature of
he body with a dry skin,

The latter are lined by secreting cells. The mouth of the duct of a sebaceous gland is seen on one side of the hair follicle in Fig. 77.

The Sebaceous Secretion is oily and semi-fluid. In healthy persons it lubricates the hairs and renders them glossy even when no " hair-oil " is used. It is also spread more or less over all the surface of the skin, and makes the cuticle less permeable by water, which in consequence does not readily wet the healthy skin, but runs off it, as " off a duck's back," though to a less marked extent.

The Skin as a Sense Organ.—Besides its functions as a protective covering and an excretory organ, the skin is of extreme importance, as being the seat of one of our most important senses—the *sense of touch.* (Chap. XXIII.)

Hygiene of the Skin.—The sebaceous secretion and the solid residue left by evaporating sweat form a solid film over the skin, which tends to choke the mouths of the sweat glands (the so-called " pores " of the skin) and impede their action. Yet these glands, minute though each is, have for their function to separate daily from the body a great amount of water* and some little urea and salines. Hence the importance of personal cleanliness. The whole skin, except that of the scalp, should be washed daily. Women cannot

Describe the structure of a sebaceous gland.
Describe the sebaceous secretion. Why is the hair of a healthy person, using no hair-oil, glossy ? Why does water run off the skin?
Enumerate the chief functions of the skin.
How are the mouths of the sweat glands apt to be choked up ?
Point out functions of these glands.
How much of the skin should be washed daily ?

---

* The sweat glands not merely carry off some water from the body, but serve also to regulate its temperature. When water evaporates from the surface of any object it abstracts heat from that object; and when perspiration evaporates it takes more heat from the skin,

well wash their hair every day, as it takes so long to dry: but there is no reason why a man should not immerse his head when he takes his bath. Except on parts of the skin especially exposed to contamination, soap should only be used occasionally—say once or twice a week; its employment is quite unnecessary for cleanliness, except on exposed parts of the body, if frequent bathing is a habit and the skin be well rubbed afterwards until dry. Soap nearly always contains an excess of alkali, which in itself injures some skins, and, besides, is apt to combine chemically with the sebaceous secretion and carry it too freely away. Persons whose skin is injured by soap, will find in cornmeal a good substitute. No doubt many folk go about in very good health with very little washing ; contact with the clothes and other external objects keeps the skin excretions from accumulating to any very great extent. But apart from the duty of personal cleanliness imposed on every one as a member of society in daily intercourse with others, the mere fact that the healthy body can manage to get along under unfavorable conditions is no reason for exposing it to them. A clogged skin throws more work on the lungs and kidneys than their fair share, and the evil consequences may be experienced any day when something else throws another extra strain upon them.

Bathing.—One object of bathing is to cleanse the skin;

How often should soap be used in the bath? Why is the too frequent use of soap not desirable? What is a good substitute for it in cases where soap is injurious to the skin? How does an unclean skin influence internal organs? When are its results apt to show themselves?

The sweat glands secrete more vigorously when the body is heated, and the evaporation of their secretion cools it. In most fevers the sweat glands are paralyzed; and the abnormally warm body is not cooled by loss of the heat, which in health would have been carried off by the evaporating sweat, but it is also useful to strengthen and invigorate the whole frame. For strong healthy persons a cold bath is the best; in severe weather the temperature of the water should be raised to 15° C. (about 60° F.), at which it still feels quite cool to the surface. The first effect of a cold bath is to contract all the skin-vessels and make the surface pallid. This is soon followed by a reaction, in which the skin becomes red and full of blood, and a glow of warmth is felt in it. The proper time to come out of the bath is while this reaction lasts, and after emersion it should be promoted by a good rub. If the stay in the cold water be too prolonged, the state of reaction passes off, the skin again becomes pallid, and the person probably feels cold, uncomfortable, and depressed all day: such bathing is injurious instead of beneficial; it lowers instead of stimulating the activities of the body. How long one may remain in cold water with benefit, depends greatly on the individual; a vigorous man can bear and set up a healthy reaction after much longer immersion than a feeble one; moreover, a person used to cold bathing can with benefit remain in the water longer than one not accustomed to it. Of course, apart from this, the temperature of the water has a great importance. Water which feels cold to the skin may, as shown by the thermometer, vary within very wide limits of temperature. The colder it is, the shorter the time which it is wise to remain in it.

When to Bathe.—It is perfectly safe to bathe when warm,

What ends are obtained by bathing? What sort of a bath should healthy persons take? What is the primary effect of a cold bath on a healthy person? What follows next? When should one leave a cold bath? What happens if one stays too long in a cold bath? Point out conditions which influence the time of remaining with benefit in a cold bath.

provided the skin is not perspiring profusely; the common belief to the contrary notwithstanding. On the other hand, no one should enter a cold bath when feeling chilly, or in a depressed vital condition. It is not wise to take a cold bath immediately after a meal, for the afterglow of the skin tends to draw away too much blood from the digestive organs, which are then actively at work. The best time for a long bath is two or three hours after breakfast; but for a brief daily dip there is no better time than while the body is still warm from bed.

Shower Baths abstract less heat from the body than an ordinary cold bath, and at the same time give it a greater stimulus, tending to set up the warm reaction. Hence they are valuable to persons in not very vigorous health.

Warm Baths, except occasionally for purposes of cleanliness, are medical remedies, and not proper things for daily use. While promoting the tendency to perspiration (which it is often important to do in disease), they also, if often repeated, lower the general vigor of the body. Persons in feeble health, who cannot stand an ordinary daily cold bath, may diminish the shock to the system by raising the temperature of the water they bathe in to any point at which it still feels cool to the skin.

Action of Alcohol on the Skin.—One of the first effects of alcohol is a flushing of the skin. The nerves of the blood-vessels that should control the passage of the blood through them are paralyzed. The blood-vessels expand and an undue amount of blood flows to the surface, caus-

Is it ever safe to bathe while warm? Point out conditions when a cold bath should be avoided. Why is it not wise to take a cool bath soon after a meal? What is the best time for a prolonged cold bath? What the best time for a brief daily dip?

Why are shower baths better than immersion baths for persons in enfeebled health?

ing a transference of heat from internal parts to the skin, in which the main organs of the temperature-sense (p. 374) are located, this produces a temporary feeling that the body is warmer; but the final result is a loss of animal heat to the air, and a decrease of the temperature of the body as a whole.

With occasional alcoholic indulgence this dilation of the cutaneous blood-vessels is temporary; with repeated, it becomes permanent. The skin is then congested and puffy, and on exposed parts it is seen to have a purplish or reddish blotched appearance; pimples appear on parts, such as the nose, where the natural circulation is more feeble. The result is the peculiar degraded look of the sot's face. The congestion interferes with the nutrition of the skin; the epidermis (p. 298) is imperfectly nourished and collects in scaly masses, interfering with the proper action of the sweat-glands, thus throwing undue work on the kidneys.

Should healthy persons take daily warm baths? What is the consequence of frequent warm baths?

How should persons in feeble health regulate the temperature of their daily bath?

Describe the effects of alcohol on the skin.

APPENDIX TO CHAPTER XIX.

To demonstrate the anatomy of the renal organs proceed as follows:

1. Kill a rat in any merciful way; placing it under a bell-jar with a sponge soaked in ether is a good method.

2. Open the abdomen of the animal, remove its alimentary canal, and cut away (with stout scissors) the ventral portion of the pelvic girdle. The dark-red *kidneys* will then be easily recognized on each side of the dorsal part of the abdominal cavity, the right one nearer the head than the left.

3. Dissect away neatly the connective tissue, etc., in front of the vertebral column, so as to clean the *inferior vena cava* and the *abdominal aorta*. Trace out the *renal arteries* and *veins*.

4. Find the *ureter*, a slender tube passing back from the kidney towards the pelvis: it leaves the inner border of the kidney behind the vein and artery; and lying, at first, at some distance from the middle line, converges towards its fellow as it passes back.

5. Follow the ureters back until they reach the *urinary bladder;* dissect away the tissues around the latter and note its form, etc.

6. Open the bladder; find the apertures of entry of the ureters, and pass bristles through them into those tubes. Note the *mucous membrane* lining the bladder.

7. Remove one kidney from the body and divide it from its outer to near its inner border; turn the two halves apart (still leaving them connected by the tissues at the inner border), and examine the cut surfaces.

8. Note at the inner border (*hilus*) the dilatation (*pelvis*) of the ureter; the outer, darker, granular *cortical portion* of the kidney, and the inner, paler, smoother *medullary portion;* the *papilla* formed by a projection of the medullary substance at the hilus, contained in an expansion (*calyx*) of the pelvis of the ureter.

9. Obtain a fresh sheep's kidney. Divide it by a section made through it from its outer to its inner border. On the cut surfaces the cortex and medulla will be more readily demonstrated than on the rat's kidney. The *pyramids of Malpighi* will also be easily seen, and the offshoots of the cortex extending between them.

CHAPTER XX.

WHY WE NEED A NERVOUS SYSTEM. ITS ANATOMY.

The Harmonious Co-operation of the Organs of the Body.—We have already learned that the body consists of a vast number of cells and fibres, combined to form organs; and that each kind of cell or fibre and each organ has its own peculiar structure, properties, and uses. Except in so far as the blood, passing from organ to organ, carries matters from one to another, and indirectly enables each organ to act upon the rest, we have as yet seen no means by which all this collection of organs is made to work together, so that each shall not merely look after itself, but regulate its activity in relation to the needs or dangers of other parts of the body.

That the organs do co-operate we all know. The lids shut when an object threatens to touch the eye, and (without our thinking about it at all) thrust themselves in the way so as to protect the more tender eyeball. When we are using the muscles of the legs vigorously the muscles of respiration hurry their action, and, consequently, oxygen is conveyed more rapidly to the blood for the supply of the working leg muscles, and the carbon dioxide produced in great quantity by these muscles is quickly removed. When the sole of the foot is tickled the muscles of the thigh and

Of what is the body made up? How do the various kinds of cells, fibres, and organs differ? How does the blood enable each organ to influence the rest?

Give an example showing the co-operation of the organs of the body. Give another example. A third.

leg, which are not directly interfered with at all, contract and jerk the foot away from its tormentor. Everywhere we find this co-operation among the organs; and it is only by such co-operation that our bodies are able to continue alive. In Æsop's fable we are told how the arms and jaws declined to work any longer in providing and grinding food for the lazy stomach, and how they soon came to grief in consequence. We might extend the fable, and go on to state how afterwards the stomach made up its mind to digest and absorb just as much food as it wanted for itself, and not bother about supplying those cantankerous arms and jaws, and the moral would be the same: if the stomach ceased to work for the other parts they soon would cease to be able to send food to it, and so it would itself starve in turn.

How a Man differs from a Collection of Living Organs. —Throughout the body, heart, lungs, stomach, intestines, liver, muscles, and skin, all need one another's aid to obtain food and oxygen, to remove wastes, and to avoid dangers. This co-operation makes the individual human being; a mere mass of living organs, arranged together in the form of man's body, but each acting without reference to the rest, would no more make a *man* than a mob of strong men would make an army. As in the mob the reckless courage of some, the personal cowardice of others, the uncontrolled ambition of a few, would make the crowd nearly useless for military purposes in spite of the merits of its individual members, so in the body; if the organs were

Is co-operation between its organs general throughout the body? Is it important? Illustrate the importance of co-operation between the parts of the body.

For what purposes do the different organs need one another's help? Is the co-operation of organs necessary to make an individual human being? Illustrate.

not disciplined, controlled, and guided, so as to work to-gether for the good of the whole, death would very soon result. As a matter of fact this is the way in which death almost always does begin. The body is not built like the deacon's "one-hoss shay," to run till every part of it gives out at the same moment. Some important organ ceases to do its part properly; as a consequence the whole complex mechanism is thrown out of gear, and death results.

Co-ordination means controlling the activities of a num-ber of working things (whether men, or organs, or ma-chines) for the attainment of a definite end. A pro-miscuous and undirected crowd of competent bricklayers, carpenters, hod-carriers, and so forth, would be quite in-competent to build a house. There might be present abundant energy and skill to construct walls and floors and roof; but if each man worked for himself and took no heed of the rest the result would be an odd building, if any at all. Hence the whole work is placed under the control of a master builder, who guides the activities of individ-uals according to the needs of the moment: undirected workmen, if conscientious, would work just as hard with-out supervision, but they would work unprofitably. The healthy body may be regarded as made up of a number of conscientious workers, the organs, who are concerned in building it and keeping it in repair, each one acting so as to co-operate with the rest for the attainment of the com-mon end. The master builder, or "boss," if we may use

How does death usually begin? What happens when some im-portant organ ceases to do its duty?

What is meant by co-ordination? Illustrate. How may the healthy body be regarded when compared with the workmen con-cerned in building a house?

such a word, is represented by the *nervous system*, which is in communication with all the other organs, is influenced by the condition and the needs of every part at each moment, and guides the activity of all the others accordingly. Part of this control is exercised consciously and with the co-operation of the " will," but much more is carried on by the nervous system without our knowing anything about it.

Nerve-Trunks and Nerve-Centres.—In dissecting the body numerous white cords are found which at first sight might be taken for tendons. That they are something else soon becomes clear, since a great many of them have no connection with muscles, and those which have, usually enter near the middle of the belly of the muscle, instead of being fixed to its ends as most tendons are. These cords are *nerve-trunks :* followed from the middle line of the body each (Fig. 79) will be found to break up into finer and finer branches, until the subdivisions become too small to be followed without the aid of a microscope. Traced towards the middle of the body the trunk will, in most cases, be found to increase by the union of others with it, and ultimately to join a much larger mass of different structure, from which other similar trunks spring. This mass is *nerve-centre.* The end of a nerve attached to the centre is, naturally, its *central,* and the other its *distal* or *peripheral end.*

What is it in the body which represents the master builder? What are the relations between the nervous system and the other organs of the body? Is the control of the nervous system always consciously exercised?

How may we recognize that certain white cords found on dissecting the body are not tendons? What are they? What is found when nerve-trunks are traced towards the outer parts of the body? What when traced in the opposite direction? What is a nerve-centre? What is meant by the peripheral end of a nerve?

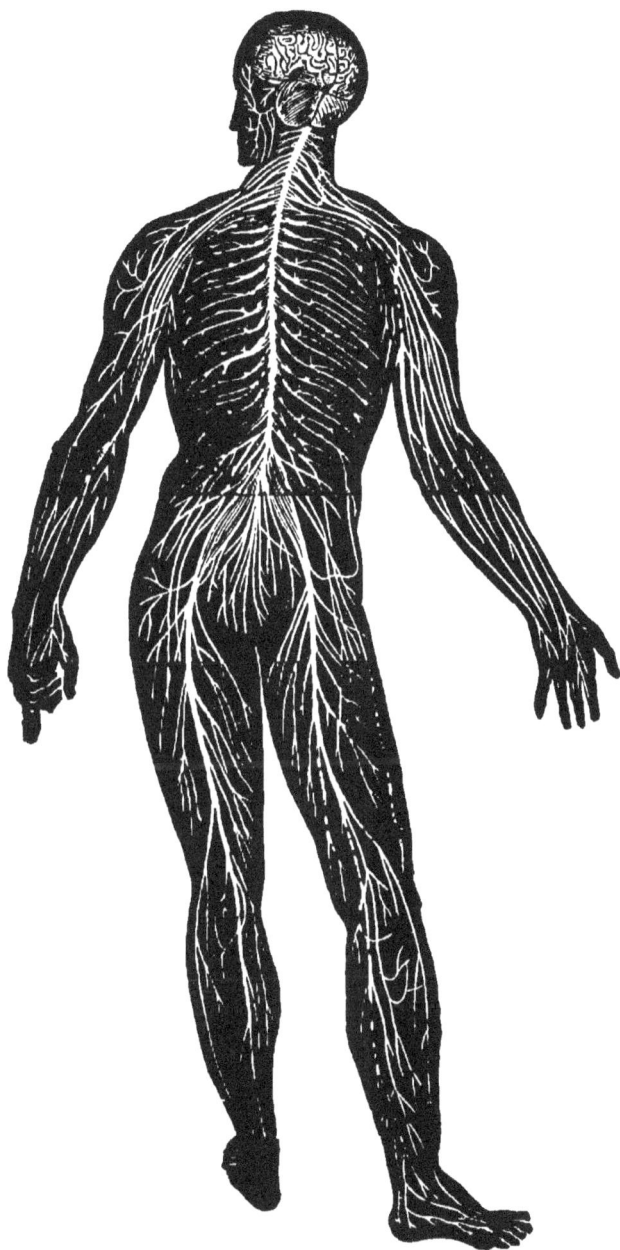

Fig. 79.—Diagram illustrating the general arrangement of the nervous system.

Nerve-centres give origin to nerve-trunks ; these radiate all over the body, branching and becoming smaller and smaller as they proceed from the centre ; finally they end in or among the cells and fibres of the various organs. The general arrangement of the nerve-centres and of the larger nerve-trunks of the body is shown in Fig. 79.

The Main Nerve-Centres.—The great majority of the nerve-trunks take their origin from the *brain* and *spinal cord,* which together form the great *cerebro-spinal centre.* Some nerves, however, commence in rounded or oval masses, which vary in size from that of the kernel of an almond down to microscopic dimensions, and which are widely distributed in the body. Each of these smaller centres is called a *ganglion.* A considerable number of the largest ganglia are united directly to one another by nerve-trunks, and give off nerves especially to blood-vessels and to the organs in the thoracic and abdominal cavities. These ganglia and their branches form the *sympathetic nervous system* (Figs. 1 and 2), as distinguished from the cerebro-spinal nervous system, consisting of the brain and spinal cord and the nerves proceeding from and to them.

The Cerebro-Spinal Centre and its Membranes.—Lying in the skull is the *brain,* and in the neural canal of the vertebral column the *spinal cord* or *spinal marrow,* the two being continuous through the *foramen magnum*

To what do nerve-centres give origin? What becomes of nerve-trunks?

From what organs do most nerves arise? What is meant by the cerebro-spinal centre? From what do those nerves arise which are not directly connected with brain and spinal cord? What are the smaller nerve-centres named? What is the sympathetic nervous system? Where is the brain placed? What organ lies in the neural canal of the backbone? Through what opening do brain and spinal cord unite?

(Fig. 20) of the occipital bone. This cerebro-spinal centre consists of similar right and left halves, incompletely separated by grooves and fissures. Brain and spinal cord are very soft and easily crushed; accordingly, both are placed in almost completely closed bony cavities, and are also enveloped by membranes which give them support. These membranes are three in number. Externally is the *dura mater*, tough and strong, and composed of connective tissue. The innermost enveloping membrane of the cerebro-spinal centre, in immediate contact with the proper nervous parts, is the *pia mater*, less dense and tough than the dura mater. Covering the outside of the pia mater is a layer of flat cells; a similar layer lines the inside of the dura mater, and these two layers are described as the third membrane

Why are brain and spinal cord placed in bony chambers? How many membranes also support them? What is the outside membrane named? What are its mechanical properties? Of what is it composed? What is the pia mater?

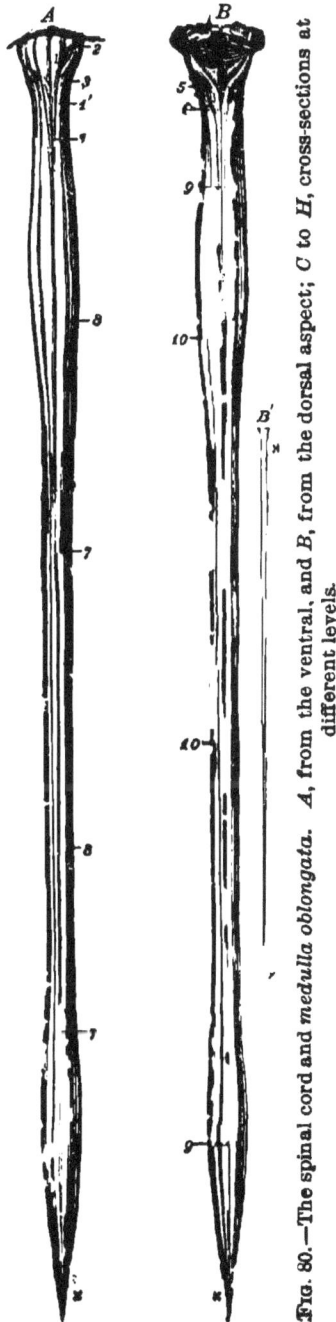

FIG. 80.—The spinal cord and medulla oblongata. *A*, from the ventral, and *B*, from the dorsal aspect; *C* to *H*, cross-sections at different levels.

of the ᵔerebro-spinal centre, called the *arachnoid*. In the space between the two layers of the arachnoid is a small quantity of watery *cerebro-spinal liquid.*

The Spinal Cord (Fig. 80) is nearly cylindrical in form, being, however, a little wider from side to side than dorso-ventrally, and tapering off at its posterior end. Its average diameter is about ¾ inch and its length 17 inches. It weighs 1½ ounces. There is no marked limit be-

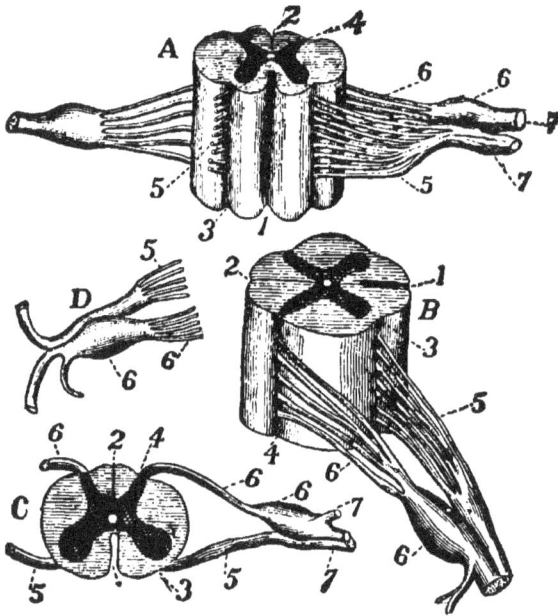

Fig. 81.—The spinal cord and nerve-roots. *A*, a small portion of the cord seen from the ventral side: *B*, the same seen laterally; *C*, a cross-section of the cord; *D*, the two roots of a spinal nerve; 1, anterior (ventral) fissure; 2, posterior (dorsal) fissure; 3, surface groove along the line of attachment of the anterior nerve-roots; 4, line of origin of the posterior roots; 5, anterior root filaments of a spinal nerve; 6, posterior root filaments; 6′, ganglion of the posterior root; 7, 7′, the first two divisions of the nerve-trunk after its formation by the union of the two roots.

What is the arachnoid? What is the cerebro-spinal liquid? What is the general form of the spinal cord? Its average diameter? Its length? Its weight? How does it connect with the brain?

tween the spinal cord and the brain, the one passing
gradually into the other. In its course the cord presents
two expansions, an upper, 10 (Fig. 80), the *cervical enlargement*, reaching from the third cervical to the first
dorsal vertebræ, and a lower or *lumbar enlargement*, 9,
opposite the last dorsal vertebræ.

Running along the middle line on both the ventral and
the dorsal aspects of the cord are fissures which (*C*, Fig.
81) nearly divide it into right and left halves.

A transverse section, *C*, shows that the substance of the
cord is not alike throughout, but that its *white* superficial
layers envelop a central *gray substance* arranged somewhat
in the form of a capital II. Each half of the gray matter
is crescent-shaped, and the crescents are turned back to
back and united across the middle line by the *gray commissure.*

The Spinal Nerves.—Thirty-one pairs of spinal nerves
join the spinal cord in the neural canal of the vertebral
column, entering through the intervertebral foramina (p.
44). Each divides in the foramen into a *dorsal* and
ventral portion, often named the *posterior* and *anterior*
roots of the nerve (6 and 5, Fig. 81), and these are attached
to the sides of the cord. On each dorsal root is a *spinal*
ganglion (6′, Fig. 81), placed where it joins the anterior or

What expansions are seen in it? Where is each placed?
How are the right and left halves of the cord separated?
Is the spinal cord alike all through? What are the colors of its
outer and of its inner portions? What is the form of the gray matter of the cord as seen on cross-sections? How are the crescents
united?
How many spinal nerves are there? What do they join? How
do they get into the neural canal? Where do they divide? What
are the divisions named? To what are they attached? What is
found on each posterior root?

ventral root to form the *common nerve-trunk*. Immediately after the mixture of fibres from both roots, the trunk begins to divide into branches for the supply of some region of the body.

The Brain (Fig. 82) is far larger than the spinal cord and more complex in structure. It weighs on the average about 50 ounces in the adult. The brain consists of three main masses, each with subsidiary parts, following one another in series from before back, and respectively known as the *fore-brain*, *mid-brain*, and *hind-brain*. In man the fore-brain, *A*, weighing about 44 ounces, is much larger than all the rest put together and laps over them.

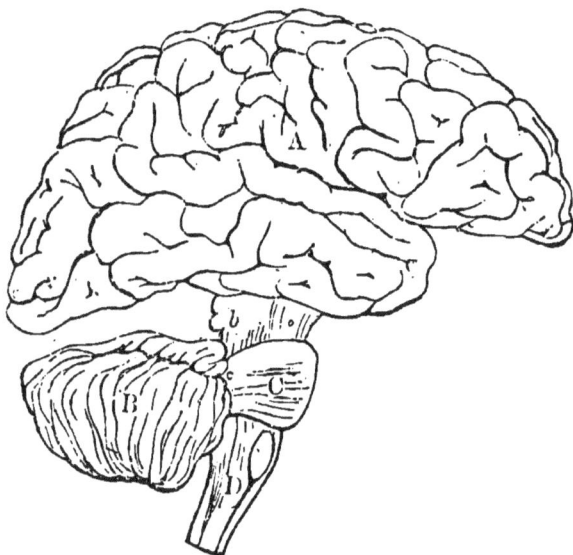

FIG. 82.—Diagram illustrating the general relationships of the parts of the brain. *A*, fore-brain; *b*, mid-brain; *B*, cerebellum; *C*, pons Varolii; *D*, medulla oblongata; *B*, *C*, and *D* together constitute the hind-brain.

What becomes of the common trunk formed by the mixture of the roots?

Point out characters in which the brain differs from the spinal cord. What is its weight? Of what main divisions is the brain composed? Which is the largest division? Its weight?

It is chiefly formed of two large convoluted masses, separated from one another by a deep fissure, and known as the *cerebral hemispheres*. The great size of these is very characteristic of the human brain. Beneath each cerebral hemisphere is an *olfactory lobe* (*I*, Fig. 84), inconspicuous in man but often larger than the cerebral hemispheres, as in most fishes. The mid-brain, *b*, forms a connecting isthmus between the two other divisions. The hind-brain consists of three main parts: on its dorsal side is the cerebellum, *B*, Fig. 82; on the under side is the *pons Varolii*, *C*, Fig. 82; and behind is the *medulla oblongata*, *D*, Fig. 82, which joins the spinal cord.

In nature the main divisions of the brain are not sepa-

Cb

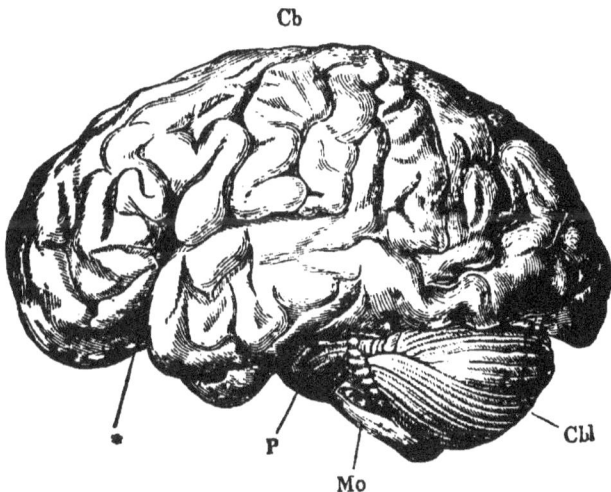

*

P

Mo

Cbl

Fig. 83.—The brain from the left side. *Cb*, the cerebral hemispheres forming the main bulk of the fore-brain; *Cbl*, the cerebellum; *Mo*, the medulla oblongata; *P*, the pons Varolii; *, the fissure of Sylvius.

By what is the fore-brain chiefly formed? What lies below the cerebral hemispheres? Are the olfactory lobes ever larger than the cerebral hemispheres? What does the mid-brain form? Name the main divisions of the hind-brain? State their relative positions, What part of the brain joins the spinal cord?

rated so much as has been represented in the diagram for the sake of clearness, but lie close together as represented in Fig. 83; and the mid-brain is entirely covered in on its dorsal side. Nearly everywhere the surface of the brain is folded, the folds, known as the *convolutions*, being deeper and more numerous in the brain of man than in that of the animals nearest allied to him zoologically.

The brain, like the spinal cord, consists of gray and white nervous matter, but somewhat differently arranged; for while the brain, like the cord, contains gray nerve-matter in its interior, a great part of its surface is also covered with it. By the external convolutions of the cerebellum and of the cerebral hemispheres the surface over which this gray substance is spread is very much increased.

The Cranial Nerves.—Twelve pairs of nerves leave the skull cavity by apertures in its base; they are known as the cranial nerves. Most of them spring from the under side of the brain, which is represented in Fig. 84. The *first pair*, or *olfactory nerves*, are the nerves of smell; they arise from the under sides of the olfactory lobes, *I*, and pass out through the roof of the nose. The *second pair*, or *optic nerves*, *II*, are the nerves of sight; they spring from the mid-brain, and, under the name of the *optic tracts*, run down to the under side of the fore-brain, where they unite

Is the surface of most of the brain smooth? What are the folds called? How does a man's brain differ, as regards its convolutions, from an ape's?

Of what does the brain consist? How does the arrangement of white and gray matter in it differ from that of the spinal cord? How is the surface on which the gray matter is spread increased?

How many cranial nerves are there? Where do most of them originate? Name the first pair. Where do they arise? Where do they pass out? Name the second pair. What are the optic tracts?

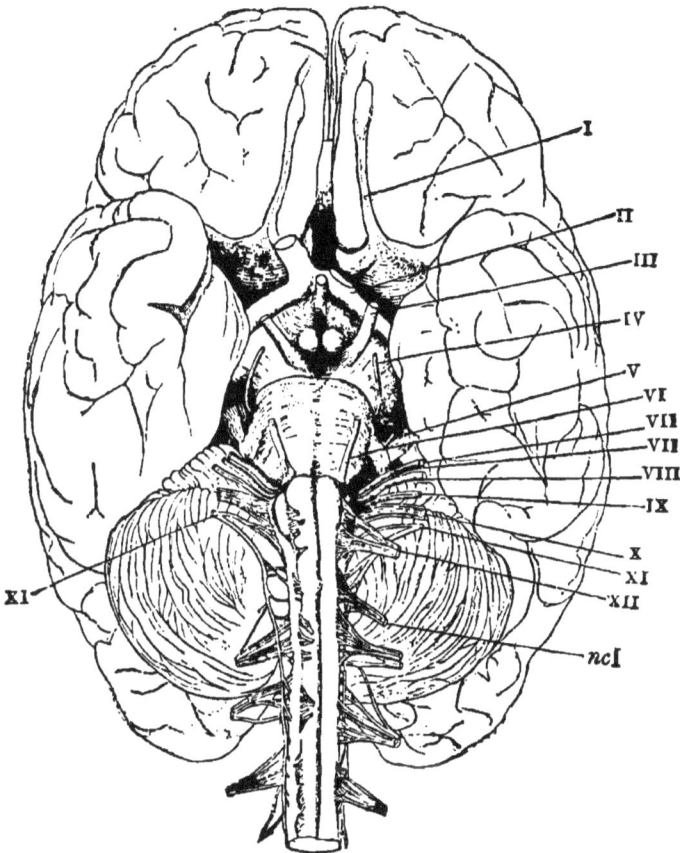

Fig 84.—The base of the brain. The cerebral hemispheres are seen over-lapping all the rest.   *I*, olfactory lobes; *II*, optic tract passing to the optic commissure from which the optic nerves proceed; *III*, the third nerve or *motor oculi; IV*, the fourth nerve or *patheticus; V*, the fifth nerve or *trigeminalis; VI*, the sixth nerve or *abducens; VII*, the seventh or facial nerve or *portio dura; VIII*, the auditory nerve or *portio mollis; IX*, the ninth or glosso-pharyngeal, *X*, the tenth or pneumogastric or *vagus; XI*, the spinal accessory; *XII*, the hypoglossal; *ncI*, the first cervical spinal nerve.

to form the *optic commissure*, from which an *optic nerve* proceeds to each eyeball.

All the remaining cranial nerves arise from the hind-brain.   The *third pair, III.*, (*motores oculi*, or *movers of the*

What is the optic commissure?   Where does each optic nerve go?
How many pairs of cranial nerves arise from the hind-brain?
Name the third pair.   What is their distribution?

eyeball,) are distributed to most of the muscles which move the eyeball and also to that which lifts the upper eyelid.

The *fourth pair, IV, (pathetici,)* are quite small; each goes to one muscle of the eyeball.

The *fifth pair* of cranial nerves, *V, (trigeminals,)* resemble the spinal nerves in having two roots, one of which possesses a ganglion (the *Gasserian ganglion*). Beyond the ganglion the two roots form a common trunk which divides into three main branches. The first of these, the *ophthalmic*, is distributed to the muscles and skin over the forehead and upper eyelid; and also gives branches to the mucous membrane lining the nose, and to the integument over that organ. The second division (*superior maxillary nerve*) of the trigeminal gives branches to the skin over the temple, to the cheek between the eyebrow and the angle of the mouth, and to the upper teeth; as well as to the mucous membrane of the nose, pharynx, soft palate and roof of the mouth. The third division (*inferior maxillary*) is the largest branch of the trigeminal. It is distributed to the side of the head and the external ear, the lower lip and lower part of the face, the mucous membrane of the mouth and the anterior two thirds of the tongue, the lower teeth, the salivary glands, and the muscles which move the lower jaw in mastication.

The *sixth pair* of cranial nerves, *VI, (abducentes,)* are distributed each to one muscle of the eyeball on its own side.

What is the distribution of the fourth pair?
Name the fifth pair of cranial nerves. How do they resemble the spinal nerves? What is the ganglion on one root called? Into how many main branches does the common trunk divide? Name the first branch. State its distribution. The second branch. Its distribution. The third main branch. Its distribution.
Name the sixth pair of cranial nerves. Where do they go?

The *seventh pair* (*facial nerves*), *VII*, are distributed to most of the muscles of the face and scalp.

The *eighth pair* (*auditory nerves*), *VIII*, are the nerves of hearing, and are distributed to the inner part of the ear.

The *ninth pair* of cranial nerves (*glosso-pharyngeal*), *IX*, are distributed chiefly to tongue and pharynx.

The *tenth pair* (*pneumogastric nerves* or *vagi*), *X*, give branches to the pharynx, gullet and stomach, the larynx, windpipe and lungs, and to the heart. The vagi run farther through the body than any other cranial nerves.

The *eleventh pair* (*spinal accessory nerves*), *XI*, do not arise mainly from the brain, but from the spinal cord by a number of roots attached to its upper portion, between the anterior and posterior roots of the proper spinal nerves. Each enters the skull cavity alongside of the spinal cord and, getting a few filaments from the medulla oblongata, passes out by the same aperture as the glosso-pharyngeal and pneumogastric nerves. Outside the skull the spinal accessory divides into two branches, one of which joins the pneumogastric trunk, while the other is distributed to muscles about the shoulders.

The *twelfth pair* of cranial nerves (*hypoglossi*), *XII*, are distributed mainly to the muscles of the tongue.

The Sympathetic Nervous System.—The ganglia which form the main centres of the sympathetic nervous system

Name the seventh pair of cranial nerves. To what parts are they distributed?

What is the eighth pair called? What are their functions? Where do they end?

Name the ninth pair. State their distribution.

Name the tenth pair. State their distribution.

Name the eleventh pair. Where do they arise? How do they get into the brain-case? With what nerves do they leave the brain-case? State their distribution.

Name the twelfth pair of cranial nerves. State their distribution.

lie in two rows (*s*, Fig. 1, and *sy*, Fig. 2), one on each side of the bodies of the vertebræ. Each ganglion is united by a nerve-trunk with the one in front of it and the one behind it, and so two chains are formed reaching from the base of the skull to the coccyx. In the trunk region these chains lie in the ventral cavity, their relative position in which is indicated by the dots *sy* in the diagrammatic transverse section represented on p. 9 in Fig. 2.

Each sympathetic ganglion is united by branches to neighboring spinal nerves, and near the skull to various cranial nerves also; from the ganglia and their uniting cords arise numerous trunks, which in the thorax and abdomen form networks, from which nerves are given off to the organs situated in those cavities. Many sympathetic nerves finally end in the walls of the blood-vessels of various organs. To the naked eye they are commonly grayer in color than the cerebro-spinal nerves.

By means of the junctions between the cranial and spinal nerves and the sympathetic system the brain is enabled to control the parts supplied by the sympathetic system.

The Three Kinds of Nerve Tissue.—The microscope shows that the nervous organs contain tissues peculiar to themselves, known as *nerve-fibres* and *nerve-cells*. The

How are the ganglia of the sympathetic system arranged?
How is each united to others? How far through the body do the chains of sympathetic ganglia extend? Where are the sympathetic chains situated in the trunk of the body?
How are the sympathetic ganglia united with spinal and cranial nerves? What arise from the ganglia? What do they form in the trunk region of the body? Where do many sympathetic nerve-fibres end? How do sympathetic nerves differ to the unaided eye from spinal or cerebral nerves? How is the brain enabled to control the parts supplied by the sympathetic system?
How many kinds of nerve-tissue are there? What are the peculiar nerve-tissues called?

cells are found in the centres only; while the fibres, of which there are two main varieties known as the *white* and the *gray*, are found in both trunks and centres; the white variety predominating in the cerebro-spinal nerves and in the white substance of the centres, and the gray in the sympathetic trunks and the gray portions of the central organs.

White Nerve-Fibres consist of extremely delicate threads, about $\frac{1}{2000}$ inch in diameter, but frequently of a length

A.

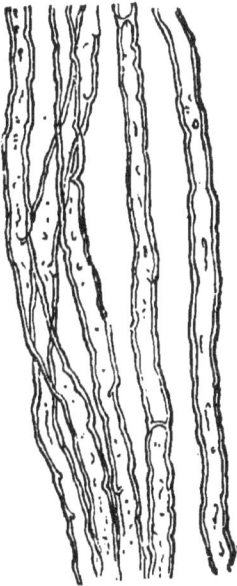

FIG. 85. FIG. 86.

FIG. 85.—White nerve-fibres soon after removal from the body and when they have acquired their double contour.

FIG. 86.—Diagram illustrating the structure of a *white* or *medullated nerve-fibre.* 1, 1, primitive sheath; 2, 2, medullary sheath; 3, axis cylinder.

Where only do we find nerve-cells? Where are nerve-fibres found? How many main kinds of nerve-fibres are there? Name them. Where do the white nerve-fibres predominate? Where the gray?

Describe white nerve-fibres.

which is in proportion very great. Each is continuous from a nerve-centre to the region in which it ends, so that the fibres, *e.g.*, which pass out from the spinal cord and run on to the skin of the toes, are three to four feet long. If a perfectly fresh white nerve-fibre be examined with the microscope it presents the appearance of a homogeneous glassy thread; but soon it acquires a characteristic double border (Fig. 85) from the coagulation of a portion of its substance, as a result of which three layers are brought into view. Outside is a thin transparent envelope (1, Fig. 86) called the *primitive sheath;* inside this is a fatty substance, 2, forming the *medullary sheath* (the coagulation of which gives the fibre its double border), and in the centre is a core, the *axis cylinder*, 3, which is the essential part of the fibre, since near its ending the primitive and medullary sheaths are frequently absent. At intervals of about $\frac{1}{25}$ inch along the fibre are found *nuclei*. These are indications of the primitive cells which by their elongation, fusion, and other modifications have built up the nerve-fibre. In the course of a nerve-trunk its fibres rarely divide; when a branch is given off some fibres merely separate from the rest, much as a skein of silk might be separated at one end into smaller bundles containing fewer threads.

Gray Nerve-Fibres have no medullary sheath, and consist merely of an axis cylinder and primitive sheath. Gray

Is each nerve-fibre continuous from centre to end? Point out nerve-fibres three or four feet long. What is the appearance under the microscope of a quite fresh white nerve-fibre? How does it soon alter? Name the layers then seen and state their relative positions. Which is the essential part of the nerve-fibre? Give a reason for your statement. What are found at intervals along the nerve-fibre? What do they indicate?
What occurs when a nerve-trunk branches?

fibres are especially abundant in the sympathetic trunks; and they alone are found in the olfactory nerve.

Nerve-Cells.—As far as our knowledge at present goes, all nerve-fibres begin as branches of nerve-cells.

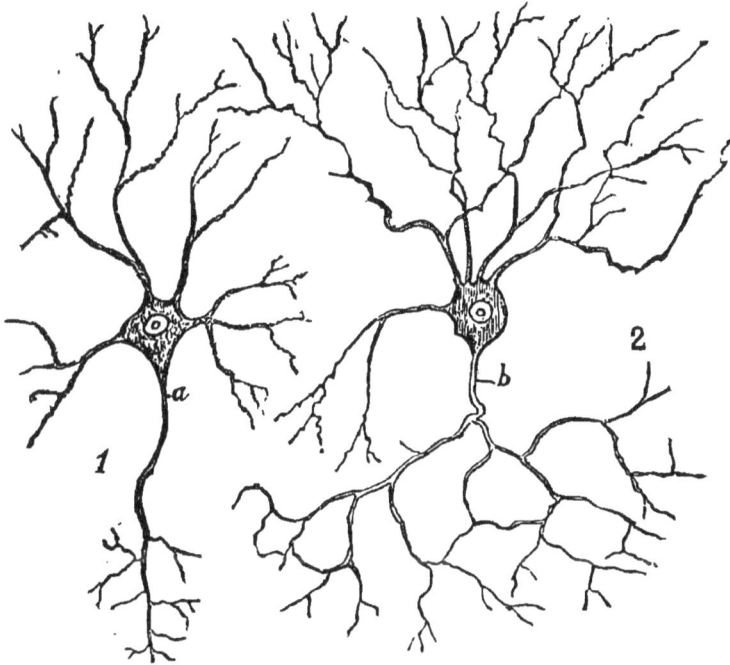

Fig. 87.—Different forms of nerve-cells from the spinal cord. 1, a cell from the anterior part of the gray matter, and believed to be connected with a motor nerve-fibre; 2, a cell from the posterior part of the gray matter, believed to be connected with sensory fibres.

At 1, Fig. 87, is shown a nerve-cell such as may be found in the anterior part of the gray matter of the spinal cord. It consists of the *cell body*, or cell protoplasm, containing a large *nucleus*, in which is a *nucleolus*. From the

Describe a gray nerve-fibre. Where are they especially numerous? Name a cranial nerve that consists entirely of them.

How do nerve-fibres begin?

Of what does a nerve-cell consist?

body of the cell arise several branches, the great majority of which rapidly subdivide. One process of the cell (*a*), although giving off several very fine branches, retains its individuality, and is continued as the axis-cylinder of a nerve-fibre in an anterior spinal root. At 2, in Fig. 87, is represented a nerve-cell from the posterior part of the gray matter of the spinal cord. It also has an axis-cylinder process, differing somewhat in its way of branching from 1, but probably ultimately giving rise to one or more cylinder-axes of sensory nerve-fibres. Cells such as those represented in **Fig. 87** are found also in many parts of the brain.

The Structure of Nerve-Centres.—These consist of white and gray nerve-fibres, of nerve-cells, and of connective tissue and blood-vessels, arranged together in different ways in the different centres.

What arise from the protoplasm of a nerve-cell? What becomes of most of the branches? How does one branch of some nerve-cells of the spinal cord differ from the remainder? As regards this branch, how do other nerve-cells differ from the above? Of what do nerve-centres consist?

APPENDIX TO CHAPTER XX.

1. The co-operation of the parts of the body may be illustrated as follows:

a. Feign a blow at a person's eye; the lids will close involuntarily, even if he be told beforehand that he is not to be actually struck.

b. Count a boy's pulse and breathing while he is sitting quietly, then let him run a hundred yards at full speed, and immediately afterwards again count pulse and breathing movements. Both will be found accelerated; the breathing, to carry off from the blood the carbon dioxide given it by the working muscles, and to bring in new oxygen to replace the large amount used by the working muscles; the heart-beat, to renew more rapidly the blood-flow through the muscles.

c. Tickle the inside of the nose with a feather. This, in itself, does not interfere with the breathing muscles; but their action will be almost at once so changed as to produce a sneeze, tending to clear and protect the nose.

2. Kill a frog with ether (note, p. 86); open its abdomen and remove the viscera. At the back of the abdominal cavity will be seen a bundle of white cords (nerve-trunks) passing back to each leg. They soon unite into one main stem (the sciatic nerve), which may be easily dissected along its course until it ends in fine branches in the hind limb.

3. Kill a frog and expose the origin of the sciatic nerve as above. With stout scissors then cut away bit by bit, and very carefully, the bodies of the vertebræ (which will be seen projecting in the middle line at the back of the abdominal cavity) until the neural canal is laid open and the spinal cord exposed. You will probably fail the first time, but on the second attempt succeed in doing this without cutting the nerve trunks as they pass between the vertebræ to join the spinal cord. On the specimen thus prepared the origin of the nerves from the spinal cord, and their division into anterior and posterior (ventral and dorsal) roots before they join the cord can be demonstrated, also the ganglionic enlargements on the posterior roots.

4. The general form, the cervical and lumbar enlargements, etc., of the spinal cord may be shown on a frog. Having killed the animal, remove the skin and muscles on the dorsal side of the spinal column. With great care cut away the upper two thirds of the neural arches of the vertebræ. Then remove the upper half of the skull cavity. Gently raising piece by piece the exposed brain and spinal cord, divide the nerves which spring from them and lift out the whole cerebro-spinal centre and place it in alcohol for twenty-four hours. Demonstrate the origin of nerves from both brain and cord, the union of the brain and cord, etc. etc. The specimen may be preserved in alcohol for future use.

5. A frog's brain differs in many important points from that of man, as in the very small cerebellum, the comparatively small cerebral hemispheres, the comparatively large mid-brain and the absence of convolutions. To demonstrate the main anatomical features of the brain that of a mammal is necessary.

a. Obtain a fresh calf's or sheep's head from a butcher. Dissect away the skin and muscles covering the cranium. Then with a small saw very carefully divide the bones in a circular direction, so as to cut off those of the crown of the head. Next carefully remove the loosened bones of the top of the skull, tearing them away from the dura mater lining them. So far the specimen may be pre-pared previous to the meeting of the class.

b. To the class demonstrate the tough dura mater enveloping the brain; then cut it away, noting the processes which it sends between the two cerebral hemispheres and between cerebellum and cerebral hemispheres. Then cut the membrane away.

c. Note its glistening inner surface, due to the arachnoid lining it; the pia mater full of blood-vessels and closely attached to the brain; the glistening arachnoid layer covering the exterior of the pia mater. Then put the specimen aside in alcohol for a day or two. This will harden the brain substance.

d. When the brain has become somewhat hardened dissect away the pia mater on one side. Show the cerebral hemispheres and their surface convolutions, the cerebellum and its foldings, the medulla oblongata beneath the cerebellum.

e. With bone forceps cut away the remainder of the sides and roof of the skull. Then raise the brain in front, and cutting through the vessels, nerves, etc., which attach it to the base of the skull, entirely remove it from the skull cavity. On it demonstrate the cerebral hemispheres (which overlap the cerebellum much less than in man), cerebellum, mid-brain, etc.

f. Attached to the base of the brain will be found the stumps of some of the cranial nerves, though most of these will have been entirely torn off unless the dissector has some technical skill. The optic commissure, with the optic tracts leading to it and the stumps of the optic nerves leading from it, will almost certainly be found.

g. Make sections across the brain in different directions to see the gray matter spread over most of its surface, and the nodules of gray matter imbedded in its interior.

CHAPTER XXI.

THE GENERAL PHYSIOLOGY OF THE NERVOUS SYSTEM.

The Properties of the Nervous System.—If one's finger unexpectedly touches a very hot object, pain is felt and the hand is suddenly snatched away ; that is to say, sensation is aroused and certain muscles are caused to contract. If, however, the nerves passing from the arm to the spinal cord have been divided, or if they have been rendered incapable of activity by disease, no such results follow. Pain is not then felt on touching the hot body nor does any movement of the limb occur ; even more, under such circumstances the strongest effort of the Will of the individual is unable to cause any movement of his hand. If, again, the nerves of the limb have connection with the spinal cord, but parts of the cord are injured higher up, between the brain and the point of junction of the nerves of the arm with the cord, then contact with the hot object may cause the hand to be snatched away, but no pain or other sensation due to the contact will be felt, nor can the will act upon the muscles of the arm, either to make them contract or to prevent their contraction. From the comparison of what happens in such cases (which have been observed again and again upon wounded or diseased persons), with what occurs in the natural condition of things, several important conclusions may be reached:

What usually results when a hot object is unexpectedly touched? Under what circumstances do these results not occur? Can the Will cause movement of the muscles of an arm whose nerves have been cut? When the arm-nerves are intact but the spinal cord is injured near the brain, what happens on touching a hot body?

1. *The feeling of pain does not reside in the burned part itself;* for it is found that applying a hot object to the skin or pinching it arouses no sensation if the nerves between the skin and the nerve-centres be diseased or divided.

2. *The hot object when the nerves are intact originates some change which, propagated along the nerves, excites a condition of the nerve-centres accompanied by a feeling,* in this particular case a painful one. This is clear from the fact that loss of sensation immediately follows division of the nerves of the limb, but does not immediately follow the injury of any of its other parts. The change propagated along the nerve-trunks and causing them to excite the nerve-centres is called a *nervous impulse.*

3. *When a nerve in the skin is excited it does not directly call forth muscular contractions;* for if so, touching the hot object would cause the limb to be moved even when the nerve had been divided high up in the arm, while, as a matter of observation and experiment, we find that no such result follows if the nerve-fibres have been cut in any part of their course from the excited, or, in physiological phrase, the *stimulated,* part to the spinal marrow. It is therefore *through the nerve-centres that the nervous impulse transmitted from the excited part of the skin is " reflected " or sent back to act upon the muscles.*

4. The preceding fact makes it probable that *nerve-fibres*

How do we know that our feeling of pain does not reside in a burned or pinched part of the skin?

What does a touched hot object originate when the nerves are healthy? What is a " nervous impulse"?

Does a skin-nerve when excited produce directly a muscular movement? Give reason for your answer. What happens to the nervous impulse transmitted from the excited part of the skin?

Is it probable that other nerve-fibres than those arising from the skin are connected with the nerve-centres ?

pass from the centre to muscles as well as from the skin to the centre. This is confirmed when we find that if the nerves of the limb be divided the Will is unable to act upon its muscles, showing that these are excited to contract through their nerves. That the nerve-fibres concerned in arousing sensation and muscular contractions are distinct, is shown also by cases of disease in which the sensibility of the limb is lost while the power of voluntarily moving it remains; and by other cases in which the opposite is seen, objects touching the hand being felt, while it cannot be moved by the Will. We conclude therefore that certain nerve-fibres when stimulated transmit something (*a nervous impulse*) to the centres, and that these, when excited by the nervous impulse conveyed to them, may radiate impulses through other nerve-fibres to distant parts, *the centre serving as a connecting link between the fibres which carry impulses from without in, and those which convey them from within out.*

5. Further we conclude *that the spinal cord can act as an intermediary between the fibres carrying in nervous impulses and those carrying them out, but that sensations cannot be aroused by impulses reaching the spinal cord only, nor has the Will its seat there ; volition and consciousness are dependent upon states of the brain.* This follows from the unconscious movements of the limb which follow stimulation of its skin after such injury to the spinal cord as prevents the transmission of nervous impulses farther on;

Point out a fact tending to prove that the muscles are normally excited to contraction through their nerves. State facts showing that the nerves of sensation and those governing the muscles are distinct. What purpose does the nerve-centre serve?

What further conclusions may we draw from the facts already considered in this chapter? Give reasons for your answer.

from the absence, in such cases, of sensation in the part whose nerves have been injured; and from the loss of the power of voluntarily causing its muscles to contract.

6. Finally, we conclude that *the spinal cord in addition to being a centre for unconscious movements serves also to transmit nervous impulses to and from the brain;* this is confirmed by the histological observation that in addition to the nerve-cells, which are the characteristic constituents of nerve-centres, it contains the simply conductive nerve-fibres, many of which pass on to the brain. In other words the spinal cord, besides containing fibres which enter it from, and pass from it to, the skin and muscles, contains many fibres which unite it to other centres.

The Functions of Nerve-Centres and Nerve-Trunks.—From what has been stated in the previous paragraphs it is clear that we may distinctly separate the nerve-trunks from the nerve-centres. The fibres serve simply to convey impulses either from without to a centre or in the opposite direction, while the centres conduct and do much more. They take heed, some consciously and some unconsciously, of the impulses carried to them by the ingoing nerve-fibres, and then send out impulses along outgoing nerve-fibres; these impulses call into action the proper organs for the safety and well-being of the body in general. The centres do not merely transmit and reflect, they also *co-ordinate.*

Classification of Nerve-Fibres.—The nerve-fibres of the body fall into two great groups corresponding to those

For what is the spinal cord a centre? What else does it do? How does histology support the belief that the spinal cord is both a nerve-centre and a conductor of nervous impulses?

What is the function of nerve-fibres? What is done by nerve-centres in addition to conducting nerve-impulses?

Into what main groups may nerve-fibres be classified?

which carry impulses to the centres and those which carry them out from the centres. The former are called *afferent* or *sensory fibres* and the latter *efferent* or *motor*.

The posterior roots of the spinal nerves contain only afferent, the anterior only efferent, nerve-fibres.

Classification of Nerve-Centres.—Nerve-centres are of three kinds: (1) *Automatic centres,* which, without being excited by the action of any sensory nerve or by the Will, originate in themselves stimuli for efferent nerves. (2) *Reflex centres,* which act quite independently of the Will and of consciousness, but are aroused by the action of a nervous impulse conveyed to them by a sensory nerve, and in turn excite one or more efferent nerves. (3) *Conscious* or *psychic centres,* whose activity is accompanied by some kind of mental action; as feeling, or willing, or reasoning.

The Psychic Nerve-Centres lie in the fore-brain, and mainly in the gray matter of its convolutions. If by any accident the cerebral hemispheres of a pigeon should be so injured as to be destroyed and all the rest of its body left intact, the animal could still control its muscles so as to execute many movements, but it would give no sign of consciousness. Left to itself it would stand still until it died; corn and drink placed before it would arouse in it no idea of eating; it would die of starvation surrounded by food. Yet it could move all its muscles, and if food should be placed in its mouth it would swallow it. If its tail should be pulled it would walk forward; if it should be put on its

What fibres are found in the posterior spinal nerve-roots? What in the anterior?

Name the main varieties of nerve-centres. What is done by automatic centres? What by reflex? What is the characteristic property of the psychic centres?

Where are the psychic nerve-centres located? What may be observed in a pigeon whose cerebral hemispheres have been destroyed?

back it would get on its feet; if it should be thrown into the air it would fly until it struck against something on which it could alight; if its feathers should be ruffled it would smooth them with its bill.

The difference between a pigeon in this state and an uninjured pigeon lies in the absence of the power of forming ideas or initiating movements. It has no thoughts, no ideas, no Will. We cannot predict what an uninjured pigeon will do under given circumstances: we can say beforehand what the pigeon with no cerebral hemispheres will do; it is a mere machine or instrument, which can be played upon. In such a pigeon the excitation of any given sensory nerve or nerves excites unconscious nerve-centres which set certain muscles at work, and the result of any one stimulus is always the same invariable movement. The animal exhibits no evidence of possessing any consciousness; it has no desires or emotions; it is like a piano which while untouched is silent, but when a given key is struck emits always the same note; the pigeon without its cerebral hemispheres stays quiet while left to itself, and responds to any one given stimulus always in one invariable and predicable way.

Functions of the Cerebellum.—The cerebellum is the great centre for co-ordinating the muscles of locomotion. Each step we take implies the action of many muscles and many thousands of muscular fibres; the actions of all must be very precisely graded as to amount, and very accurately arranged as to proper sequence. We do not, however, consciously think about the muscles to be used in every move-

How does such a pigeon differ from an uninjured pigeon?
What is the main function of the cerebellum? What is implied in each step that we take?

ment of each step; if we do think at all about our walking
the cerebral hemispheres simply send a message to the cerebellum, and leave it (with the aid of the spinal cord) to
regulate all the details. When we walk without thinking
about it, the contact of the foot with the ground stimulates
sensory nerves of the sole, which then stimulate the locomotor
centres; these centres excite in proper order the nerves
which control the muscles; and the co-ordinated action of
the muscles produces the next step.

A pigeon with its cerebellum destroyed and all the rest
of its nervous system intact would stand unsteadily, would
stagger when it attempted to walk, and flutter uselessly
when thrown into the air. But, having its cerebrum, it
could will and feel; it would not stand quiet until touched;
it would initiate movements when left to itself, though it
could not perform them properly. It would will, and feel,
and think, but could not co-ordinate the action of its muscles except for some simple movements, regulated by the
medulla oblongata or the spinal cord.

Automatic Nerve-Centres send out nervous impulses
through efferent nerves without waiting to be excited by
afferent nerves or by the Will. The most conspicuous are
the small nerve-centres buried in the heart, which excite its
beat even when it is separated from all the rest of the body.
Another automatic centre is that which lies in the medulla
oblongata and stimulates the nerves which control the

Do we have to think about using each muscle concerned in walking? What happens when we think about our walking? What when
we walk without thinking about it ?

What do we see in a pigeon whose cerebellum has been destroyed?
Does it initiate movements? Does it execute them well? What is
its condition as regards willing, feeling, and thinking? What can
it not do?

What is done by automatic nerve-centres? Give examples,

muscles of respiration. When this centre is cut off from all sensory nerves it still acts, and its activity goes on even against the Will. We can voluntarily hold the breath for a short time, but not long enough to kill ourselves by suffocation. Although automatic nerve-centres act independently of impulses carried to them by nerve-trunks, they are nevertheless usually more or less subject to control by them. For example, stimulating the branch of the pneumogastric nerve (p. 323) which goes to the automatic heart nerve-centres slows the beat of the organ; and a dash of cold water on the skin makes us draw a deep breath.

Reflex Centres are aroused to activity by nervous impulses conveyed to them through afferent nerves: they then excite efferent nerves and produce a movement or a secretion. Such nerve-centres do all the routine of the administrative control of the organs of the body, without troubling the psychic centres. They frequently act without the intervention of consciousness at all, and often in spite of the Will. When sugar is placed in the mouth it excites its sensory nerves ; these stimulate a centre from which nerves go to the salivary glands, and these nerves, aroused by the centre, make the gland-cells secrete and pour saliva into the mouth; no effort of the Will can stop this *reflex action,* so called because a nervous impulse sent to a centre by one set of nerve-fibres is turned back or *reflected* from it along another set. When a morsel of food enters the pharynx it excites the sensory nerves of the mucous

Can a man commit suicide by holding his breath? Are the automatic centres entirely free from control? Illustrate.

How are reflex centres excited? What is the consequence of their stimulation? What sort of work in the body is executed by reflex nerve-centres? Are we always conscious of their action? Can the Will always control them? What happens when sugar is placed on the tongue? Why is it called a "reflex action"?

membrane; these arouse a reflex centre of swallowing, which sends out nervous impulses to the swallowing muscles, and the food is sent on into the gullet whether we wish it or not. Sneezing when something irritates the mucous membrane of the nose, and coughing where some foreign mass enters the larynx, are other instances of reflex actions.

The Use of Automatic and Reflex Centres is to relieve the thinking centres of the vast amount of work which would be thrown upon them if every action of the body each moment had to be planned and willed. Were not the unconscious regulating nerve-centres always at work the mind would be overburdened by the mass of business which it would have to look after every minute. No time would be left for intellectual development if we had to think about and to *will* each heart-beat, each inspiration and expiration, and the swallowing of each mouthful of food. Moreover, during sleep, so necessary for the rest and repair of the psychic centres, the automatic and reflex centres carry on the actions essential for the nutrition of the body and the maintenance of life. If we had to reason concerning each beat of the heart and decide if it was time for it to occur and what force it should have, and then to make up our minds whether to will it or not, we could never sleep.

Habits are Acquired Reflex Actions, distinguished from *primary* or those born with us, such as sneezing, coughing, and winking. Every time a nerve-centre acts in a given way it tends to more easily act in that manner again; as a result

Give other examples of reflex actions.
What is the main use of the automatic and reflex nerve-centres?
What would result if the unconscious nerve-centres were not always at work?
What are habits? What happens when a nerve-centre acts in a given manner? What is the result?

many actions which are at first only performed with trouble and thought are after a time executed easily and unconsciously. The act of walking is a good instance; each of us in infancy learned to walk with much pains and care, thinking about each step. But the more we walked the closer became ingrained in the nervous system the connection between the stimulation of nerves in the sole when a foot touched the ground, and the sending out by the reflex nerve-centres with which they were in connection, of impulses to those muscles which had to make the next step. At last the contact of the foot with the ground, stimulating some sensory nerves, acts so readily on the "nerve-centres of walking" that the cerebral hemispheres need take no heed about it: we walk ahead while thinking of something else. In other words we have acquired a reflex action not born in us. Other instances will readily come to mind: as the difficulty with which we learned to ride, or swim, or skate, thinking about and *willing* each movement; and the ease with which we do all these things after a little practice. The trained lower nerve-centres then do all the co-ordinating work and the Will has no more need to trouble about the matter. A habit simply means that the unconscious parts of the nervous system have been trained to do certain things under given conditions, and can only be restrained from doing them by a special effort of the conscious Will. A practised rider will keep his seat unconsciously under all ordinary circumstances, and can only fall off his horse by taking some trouble to do so, by *willing* it in fact; an unskilled rider, on the other hand, must exert all his attention to avoid falling. So with what in

Illustrate by the act of walking. Give other examples. What does a "formed habit" really mean? Illustrate.

every-day language are called "habits": once we have repeated an action so often that our bodies almost unconsciously
do it, it becomes a *habit*, and needs special exercise of Will
to deviate from it. We thus find, in the tendency of the
nervous system to go on doing what it has been trained to
do, a physiological reason for endeavoring to form good
and to avoid bad habits of whatever sort, physiological,
business, social, or moral. Every thought, every action,
leaves in the nervous system its result for good or ill. The
more often we yield to temptation the stronger effort of
the Will is required to resist it. The knowledge that every
weak yielding degrades our nerve-organs and leaves its trail
in the brain, through whose action man is the "para-
gon of animals," while every resistance makes less close the
bond between the feeling and the act for all future time,
ought surely to "give us pause"; on the other hand, every
resistance of temptation helps to make subsequent resistance easier.

Hygiene of the Brain.—The brain, like the muscles, is
improved and strengthened by exercise and injured by overwork or idleness; and just as a man may specially develop
one set of muscles and neglect the rest until they degenerate,
so he may do with his brain; developing one set of intellectual faculties and leaving the rest to lie fallow until, at
last, he almost loses the power of using them at all. The
fierceness of the battle of life nowadays especially tends to
produce such lopsided mental development. How often

What happens when we have very frequently repeated an action?
Point out why it is desirable, even on physiological grounds, to form
good habits. How does every thought or act influence the nervous
system? What is the consequence of yielding to temptation? What
of resisting it?

does one meet the business man, so absorbed in money-getting that he has lost all power of appreciating any but the lower sensuous pleasures; the intellectual joys of art, science, and literature have no charm for him; he is a mere money-making machine. One, also, not unfrequently meets the scientific man with no appreciation of art or literature; and literary men utterly incapable of sympathy with science. A good collegiate education in early life, on a broad basis of mathematics, literature, and natural science, is the best security against such deformed mental growth.

The Primary Effects of a Moderate Dose of Diluted Alcohol, as a glass of whiskey and water, on one unaccustomed to it, are to cause temporary congestion of the stomach ; dilatation of blood-vessels of the skin, indicated by the flushed face ; a more rapid beat of the heart ;* nervous excitement, exhibited by restlessness and talkativeness. Then some incoherence of ideas, and often giddiness. Finally there is a tendency to sleep. On awaking the person has some feeling of depression, not much appetite, and is in general a little out of sorts for a day.

Moral Deterioration produced by Alcohol.—One result of a single dose of alcohol is that the control of the Will over the actions and emotions is temporarily enfeebled; the slightly tipsy man laughs and talks loudly, says and does rash things, is enraged or delighted without due cause.

If the dose be larger, the stage of giddiness is accompanied by diminution of the sensibility of the skin; and imperfect control over the voluntary muscles, indicated by

Name important organs to which alcohol is ultimately carried in the blood.

---

* It is doubtful if chemically pure alcohol diluted with water quickens the pulse; most ordinary alcoholic beverages, however, undoubtedly do.

defective articulation and a staggering gait. The muscles moving the eyeballs cease to work in harmony. Normally they act unconsciously, turning the eyes so that images of objects looked at are focussed on corresponding points of the retinas; and objects are seen single. Soon after the voluntary movements are affected the involuntary regulation of the eye-muscles is impaired, and objects are seen double; the eyeballs being no longer so turned as to bring images on corresponding retinal points. The stomach may also be so irritated as to lead to vomiting. Then comes deep drunken sleep; followed by headache, loss of appetite, and prostration similar to, but more marked than, that occurring after the smaller dose.

If the alcoholic indulgence be repeated, day after day, some of the above-described primary consequences become less marked; but they give way to more serious functional and structural diseases.

If the amount of alcohol be increased, further diminution of will-power is indicated by loss of control over the muscles. Habitual drinking of alcohol results in permanent over-excitement of the emotional nature and enfeeblement of the Will; the man's highly emotional state exposes him to special temptation to excesses of all kinds, and his weakened Will decreases the power of resistance: the final outcome is a degraded moral condition. He who was prompt in the performance of duty begins to shirk that which is irksome; energy gives place to indifference, truthfulness to lying, integrity to dishonesty: for even with the best inten-

Describe the primary effects of a moderate dose of dilute alcohol. Of a larger but not fatal dose.

Describe the action of alcohol on the sense-organs. On the brain and spinal cord, and the resulting diseases.

Describe the influence of alcohol on will-power.

tions in making promises or pledges there is no strength of Will to keep them. In forfeiting the respect of others respect for self is lost, and character is overthrown. Swift and swifter is now the downward progress. A mere sot, the man becomes regardless of every duty, and even incapacitated for any which momentary shame may make him desire to perform.

These results are due to the weakness of the drinker, caused by the nature of the drink. Alcohol is a brain poison; one dose of it leads to a craving for more. Happily not every beginner becomes a drunkard; but experience shows that the danger is great: no one can foretell how quickly his will-power may be weakened by alcohol.

For the inebriate there is but one hope—confinement in an asylum, where, if not too late, the diseased craving for drink may be gradually overcome, the prostrated Will regain its ascendency, and the *man* at last gain the victory over the *brute*.

Heredity.—Alcohol is a potent agent in leading to the transmission of disease, especially nervous disease, from parent to child. The drunken parent begets no healthy child. This inherited influence manifests itself in various ways. It may transmit an appetite for strong drink to the children, and the drunkard by inheritance is a more helpless slave than his father or mother. But the inherited effects are not confined to the propagation of drunkards only. They produce insanity, idiocy, epilepsy, and other affections of the brain and nervous system in the children of the drunkard, and they in turn transmit predisposition to these diseases to their child. "When alcoholism does not produce in-

State some frequent effects on the offspring of alcoholic indulgence by the parent.

sanity, idiocy, or epilepsy, it weakens the conscience, impairs the will, and makes the individual the creature of impulse and not of reason."

The Sense-organs are also affected: their acuteness of perception is dulled; and many physicians believe that *cataract* and retinal disease may be produced by drinking. The red congested white of the eye of topers is well known.

The Brain and Spinal Cord are kept in a chronic state of over-excitement. This results in inflammatory disease (*delirium tremens*); later, in fibrous degeneration, leading to certain forms of paralysis or to *epilepsy*, of which there is one variety recognized by physicians as due to alcohol.

Delirium Tremens.—Repeated drunkenness usually ends in an attack of *delirium tremens*, but this disease is more frequently the result of prolonged drinking which has never culminated in actual drunkenness. It is especially apt to occur in "those who drink hard, but keep from actual loss of consciousness, especially those engaged in hard mental work or subjected to great moral strain or shock; and, too, those of certain temperaments are peculiarly liable to it."

Few persons die in their first attack of delirium tremens, but it is nature's unmistakable warning to the tippler: let him not disregard it, unless he is prepared to die without hope, in maniacal imaginings so frightful that those around his death-bed can never recall the scene without horror !

It is preceded, usually, by loss of sleep, ideas of persecution or injury, with no foundation in fact, and slight hallucinations, especially at night; the man, meanwhile, in the day looking anxious, slightly excited, nervous and tremulous, and perhaps narrating as actual occurrences the hal-

Action of alcohol on sense organs? On brain and spinal cord?
Under what conditions may delirium tremens occur?

lucinations of the preceding night. Then the senses are partly lost; he sees spectres, horrible and foul creatures about him; has all sorts of painful, terrifying visions (whence the common name of the 'horrors'); is extremely tremulous, and either excited or lies prostrate, trembling violently on movement, sleepless, anxious, and a prey to spectres and terrors of the imagination." *

Dipsomania is often confounded with delirium tremens, but though it may lead to that disease it is an essentially different pathological state. The word properly means a mental disease in which there is periodically an irresistible passion for alcohol; in any form, no matter how distasteful, the dipsomaniac will swallow it with avidity. The disease is sometimes produced by indulgence in drink, but is more often inherited, especially from parents addicted to alcoholic excess. In the families of such, one child is often epileptic, another idiotic, a third eccentric or perhaps quite mad, and a fourth a dipsomaniac. When the fit seizes him the dipsomaniac is as irresponsible as a raving madman. His only safeguard against a frightful debauch is to place himself under restraint as soon as he perceives the symptoms which he has learned to recognize as premonitory of his fit of madness. After a time the paroxysm passes off; the patient regains self-control, loses his passion for drink, is greatly ashamed of himself if he has indulged it, and usually behaves in an irreproachable manner for some weeks or months.

What symptoms usually precede this disease? Describe the condition of a person suffering from delirium tremens. What is dipsomania?

---

* Dr. Greenfield, on "Alcohol : its Use and Abuse."

APPENDIX TO CHAPTER XXI.

1. Place a frog on the table: note that it sits up and breathes (as shown by the movements of its throat), and either stays still or jumps around as it pleases; *i.e.*, *it has a Will of its own*, and its actions cannot be predicted.

2. Etherize two frogs (note, p. 86), removing them from the etherized water the moment they become insensible. With strong sharp scissors cut off from one frog (*a*) all the head in front of the anterior margins of the tympanic membranes. in the other frog (*b*) remove the head along a line joining the posterior borders of the tympanic membranes. Place both frogs aside on a dish containing a little water for half an hour. The quantity of water should be such that while keeping the frogs moist it will not reach to the wounds.

3. The frog (*a*) which has lost its cerebral hemispheres, but retained its mid brain, cerebellum, and medulla oblongata, will be found after the above-stated time sitting up in a natural position, and breathing. Left to itself it will, however, never walk or jump; it shows no sign of possessing a will. Its heart continues to beat and its respiratory muscles to contract, but left alone it stays where it is. Turned upon its back it will regain its feet; and put into water it will swim: its muscles, and the nerves controlling them, are, therefore, quite able to act. The animal stays still not because the parts of its body necessary to produce movements are injured, but because it can no longer *will* a movement. Such a frog shows very well the dependence of *volition* upon the presence of the cerebral hemispheres.

4. The frog (*b*) will have had its whole brain removed. Its heart will continue to beat, but its breathing movements will cease, because the respiratory centre, which lies in the medulla oblongata, has been cut away. It will also lie down squat, instead of sitting up like a normal frog, because its most important muscle co-ordinating centres have been removed with the mid-brain and cerebellum. Left to itself the animal will, within half an hour of the removal of the head, pull up its hind legs into their natural position, but after this it will make no movement. It has no volition.

5. Such a frog can, however, perform many co-ordinated *reflex actions*, which may be illustrated as follows: (*a*) Pinch a toe; it will be pulled away. (*b*) Soak some blotting-paper in vinegar, and then cut the paper into small pieces about ⅓ inch square. Put these bits of paper on different regions of the frog's skin, dipping the animal in clean water after each application, to wash away the vinegar.

It will be found that the brainless creature moves its limbs so as to wipe away the acid paper placed on its skin. The frog without its brain has no Will and no consciousness; but its spinal cord when excited by afferent nerves, whose ends the vinegar stimulates, excites in turn efferent nerves which stimulate muscles, whose contraction produces a movement calculated to rub away the irritating object.

6. Now run a stout pin down the frog's neural canal so as to destroy its spinal cord. It will be found that no subsequent pinching of the creature or putting of vinegar on its skin causes any movement. Its muscles and nerve-trunks are intact, but the *spinal reflex centre*, which in the previous experiments was excited by afferent nerves, and then in turn stimulated efferent nerves, is destroyed. The heart continues to beat, on account of the automatic nerve-centres in it; but no voluntary and no reflex actions are exhibited by the animal.

7. The nerve-trunks and the muscles are, however, still active Turn the frog on its back and carefully expose (p. 331) the origins of the sciatic nerves. On pinching these, the muscles of the leg will be seen to contract. The irritation of the nerve by the pinch starts in it a nervous impulse which travels down the nerve-branches to the muscles.

CHAPTER XXII.

THE INFLUENCE UPON THE HUMAN BODY OF OPIUM, CHLORAL, AND SOME OTHER COMMONLY USED NARCOTICS.

Opium and Morphia.—Opium is a gummy mixture containing several active principles, of which the most important is morphia. The forms in which it is most frequently employed are (1) *gum opium*, the crude substance, often put up in the form of pills; (2) *laudanum*, an alcoholic extract of the gum; (3) *paregoric*, a liquid containing several substances, of which opium is the most important; (4) *morphia* and its compounds.

The Opium Habit.—Opium is perhaps the most valuable drug at the disposal of the physician. On the other hand, it is one of the most injurious substances used by mankind. It may be that it does not do so much harm in the United States as alcoholic drinks, but only because not so many persons have acquired the craving for it. Used constantly it is as certainly fatal and the habit is perhaps even harder to break; for it may be indulged more secretly and its effects are not so readily recognized. Many a one of highest gifts and noblest character has gone under in the insidious maëlstrom spread by opium for its victims. Using the drug at first as prescribed for the relief of suffering, he (or she, for more women than men are addicted to

What is opium? In what forms is it most often used?
Compare the damage done in the United States by indulgence in alcohol and opium.

opium habit) is scarcely conscious of danger before being swept on to destruction. Most medical men now fully recognize the danger, and only order prolonged use of opium with great caution. Nevertheless there are so many persons who habitually use opium that it is important to point out the disastrous results.

The Diseased Conditions produced by Continued Abuse of Opium are fairly uniform. The first phenomenon is deadening of sensibility, accompanied by mental exaltation if the dose be small. This is succeeded by unnatural sleep, disturbed by fantastic dreams.

On awaking there is great depression of mind and body: often associated with defective memory, and a feeling that something terrible is about to happen. There is muscular weakness; distaste for food, without actual nausea ; and an almost irresistible craving for another dose.

If the habit be continued further, mental and physical changes occur. Distaste and inaptitude for any kind of exertion; greatly impaired digestion ; deficient secretion of bile; sluggishness of the muscles of the intestines, causing constipation. The muscles waste, the skin shrivels, and the person looks prematurely aged. The pulse is quick, the body feverish; the eye dull, except just after the drug has been taken.

The final result is failure of the nervous system. Incomplete paralysis of the lower limbs is followed by a similar state of the muscles of the back. The victim crawls along,

Why is opium more disastrous from one point of view ?

What are the first phenomena following a dose of opium ? What is the condition of the person on awaking ?

What results follow continuance of the habit ?

What is the final result ?

bent like an old man. Death finally results from starvation, due to complete failure of the digestive organs.

Morphia.—When morphia is used, a solution of it is often injected under the skin by a fine syringe. Prolonged use of it in this way is followed by all the symptoms of chronic opium-poisoning above described. The digestive organs are not, however, as soon attacked; but the punctures of the skin repeated for weeks, several times a day, cause inflammation and ulceration.

Danger of Administering Opiates to Children.—Children are remarkably sensitive to opium and all preparations containing it. *Opiates should never be administered to children except by order of a physician.* Many an infant has been poisoned by a few drops of paregoric or of some soothing syrup given by parent or nurse to check diarrhœa or produce sleep.

Chloral.—The chloral habit is in this country at present more common than the opium habit, and, like it, more frequent among women than men.

Chloral was, on its discovery a few years ago, heralded as a wonderfully safe and certain promoter of sleep and alleviator of pain. Medical men have since learned that it is by no means so harmless a drug as they once believed; but the general public do not seem to have had their eyes opened to its danger. A great many preparations of it have been put on the market, and are sold in drug-stores to all-comers. The result is that many persons who would

In what do the consequences of injection of morphia beneath the skin resemble and differ from those of opiates taken by the mouth?

How are children peculiar as regards opiates? Under what conditions only should they be administered to children? What often results from giving opiates to infants?

Compare the frequency of the chloral and opium habits in the United States.

he itate to take opium without medical advice use chloral, believing it harmless.

Chloral, taken habitually, is at least as mischievous as opium. It should be forbidden by law to retail it in any form except on the prescription of a physician.

The chloral habit is acquired with great ease, and is very hard to break. The first phenomena of chloral disease (*chloralism*) are these: The digestion is greatly impaired; the tongue is dry and furred; there is nausea; sometimes vomiting, and a constant feeling of oppression from wind on the stomach.

Next, nervous and circulatory disturbances occur. The temper becomes irritable, the Will weak; the hands and legs tremulous; the heart-beat irregular; the face easily flushed. Sleep becomes impossible without use of the drug: when obtained it is troubled, and the person awakes unrested.

In later stages the blood is seriously altered. Its coloring matter is dissolved, and soaks through the walls of the capillary vessels, causing purplish patches on the skin. Jaundice also frequently occurs.

If the chloral-taking be still continued, death results from impoverished blood, weakened heart, or paralysis of the nervous system. Not unfrequently chloral-takers unintentionally commit suicide by indulging in too large a dose.

Tobacco contains an active principle, *nicotin*, which in its pure form is a powerful poison, paralyzing the heart.

What have medical men lately learned about chloral? Why do so many people take chloral without medical advice?

Describe the first symptoms of chloralism.

What are the symptoms in more advanced chloralism? What in the latest stages?

How does death from chloral occur?

When tobacco is smoked some of the nicotin is burned, but there are developed certain acrid vapors which have an irritant action on the mouth and throat. The effects of smoking are thus in part general, due to absorbed nicotin ; and in part local, due to irritant matters in the smoke. They vary much with the constitution, habits, and age of the smoker. *Tobacco is specially injurious to young persons whose physical development is not completed.*

The **Local Action of Tobacco** is at first manifested by increased flow of saliva. This usually passes off after some practice in smoking; dryness of the mouth follows, and consequent thirst, often leading to alcoholic indulgence; and in this, perhaps, lies the greatest danger from tobacco. The habitual smoker usually suffers eventually from what is known to medical men as "smoker's sore-throat." The inflammation often extends to the larynx, injuring the voice and producing a hacking cough, or may spread up the Eustachian tubes (p. 373) and impair the hearing. Cigarettes are especially apt to cause these symptoms. Cure is impossible unless smoking be given up. Those who draw the smoke into their lungs often suffer from chronic inflammation of the bronchial tubes in consequence.

The **General Action of Tobacco.**—The more common effects of absorption of tobacco products are to interfere with development of the red blood-corpuscles, leading to pallor and feebleness; to impair the appetite and weaken digestion; to affect the eyes, rendering the retina less sensitive; to

What is nicotin? What other injurious substances are found in tobacco-smoke? What general rule may be stated concerning the action of tobacco on the human body?

Describe the local actions of tobacco. How may tobacco-smoking injure the voice? How the hearing?

Describe the general action of tobacco on the body.

cause **palpitation** of the heart and enfeeblement of that organ; **to** induce a lassitude and indisposition to exertion that, in view of the heavy odds man has to contend with in the life-struggle, may prove the handicap that causes his failure. If success in life be an aim worth striving for, it is surely unwise to shackle one's self with a habit which cannot promote and may seriously jeopardize it.

CHAPTER XXIII.

THE SENSES.

Common Sensation and Special Senses.—Changes in many parts of our bodies are accompanied or followed by states of consciousness which we call *sensations*. All such parts (*sensitive parts*) are in connection, direct or indirect, with the brain by sensory nerve-fibres. Since all feeling is lost in any region of the body when this connecting path is severed, it is clear that all sensations, whatever their primary exciting cause, are finally dependent on conditions of the brain. Since all nerves lie within the body as circumscribed by the skin, one might be inclined to suppose that the cause of all sensations would appear to be within our bodies themselves; that the *thing felt* would be recognized as a modification of some portion of the *person feeling*. This is the case with regard to many sensations: a headache, toothache, or earache gives us no idea of any external object ; it merely suggests to each one a particular state of a sensitive portion of himself. As regards·many sensations this is not so; they suggest to us external causes, to properties of which, and not to states of our bodies, we ascribe them; and so they lead us to the conception of an external universe in which we live. A knife laid on the skin produces changes in it which

With what are all sensitive parts of the body in connection ? By what nerve-fibres? How do we know that all sensations finally depend on the brain?

Why might we suppose that the causes of all sensations would seem to lie within the body? Name sensations merely suggesting to us a state of the body itself. What do some other sensations suggest to us? Illustrate,

lead us to think not of a state of the skin, but of properties of some object outside the skin; we believe we feel a
cold heavy hard thing which is not the skin. We have,
however, no sensory nerves going into the knife and inform-
ing us directly of its condition; what we really feel are the
modifications of the body produced by the knife, although
we irresistibly think of them as properties of the knife—of
some object that is no part of the body. Let now the knife
cut through the skin; we feel no more *knife*, but experience *pain*, which we think of as a condition of ourselves. We do not say the knife is painful, but that the
finger is, and yet we have, so far as sensation goes, as much
reason to call the knife painful as cold. Applied one way
it produced local changes in the skin arousing a sensation of
cold, and in another local changes causing a sensation of pain.
Nevertheless in the one case we speak of the cold as being
in the knife, and in the other of the pain as being in the
finger.

Sensitive parts, such as the surface of the skin, through
which we get, or believe we get, information about outer
things, are of far more intellectual value to us than sensitive parts, such as the subcutaneous tissue into which the
knife may cut, which only give us sensations referred to
conditions of our own bodies. The former are called *Organs
of Special Sense;* the latter are parts endowed with *Common
Sensation.*

Common Sensations are quite numerous; for example,
pain, hunger, nausea, thirst, satiety, and fatigue.

What is meant by "organs of some special sense"? What by
parts endowed with common sensation?
Name some *common sensations.*

Hunger and Thirst.—These sensations regulate the taking of food. Local conditions play a part in their production, but general states of the body are also concerned.

Hunger in its first stages is due to a condition of the gastric mucous membrane which comes on when the stomach has been empty some time; it may then be temporarily stilled by filling the stomach with indigestible substances. But soon the feeling comes back intensified and can only be allayed by the ingestion of nutritive materials; provided these are absorbed and reach the blood their mode of entry is unessential; hunger may be stayed by injections of food into the intestine as completely as by filling the stomach with it.

Similarly, thirst may be temporarily relieved by moistening the throat without swallowing, but then soon returns; while it may be permanently relieved by water injections into the veins, without wetting the throat at all.

Both sensations depend in part on local conditions of sensory nerves, but may be more powerfully excited by poverty of the blood in foods or water; this deficiency directly stimulates the hunger and thirst centres of the brain.

The Special Senses are commonly described as five in number, but there are at least six; namely, sight, hearing, touch, the temperature sense, smell, and taste.

To what is the first stage of hunger due? How may it be temporarily stayed? Need food enter the stomach in order to alleviate hunger?

How may thirst be temporarily relieved? How permanently without swallowing water? On what do the sensations of hunger and thirst in part depend? How may they be more powerfully excited? Enumerate the special senses.

The Visual Apparatus consists of nervous tissues immediately concerned in giving rise to sensations, supported, protected, and nourished by other parts. Its essential parts are, (1) *the retina*, a thin membrane lying in the eyeball and containing microscopic elements which are so acted upon by light as to stimulate (2) *the optic nerve;* this nerve ends (3) in a part of the brain (*visual centre*) which when stimulated arouses in our consciousness a feeling or *sensation of sight.* The visual centre may be excited in very many ways, and quite independently of the optic nerve or the retina; as is frequently seen in delirious persons, in whom inflammation or congestion of the brain excites directly the visual centre and gives rise to visual hallucinations.

Usually, however, the cerebral visual centre is only excited through the optic nerve, and the optic nerve only by light acting upon the retina. The eyeball, containing the retina, is so constructed that light can enter it, and so placed and protected in the body that as a general thing no other form of energy can act upon it so as to stimulate the retina. Under exceptional circumstances we may have sight-sensations when no light reaches the eye; anything which stimulates the retina, so long as it is connected by the optic nerve with the cerebral visual centre, will cause a sight-sensation. A severe blow on the eye, even in complete darkness, will cause the sensation of a flash of light; the compression of the eyeball excites the retina, the retina excites the optic nerve,

Of what does the visual apparatus consist? What are its essential parts? What happens when the visual centre is stimulated? Is it only stimulated by the agency of light? Illustrate.

How is the visual centre usually excited? Why is light the form of energy which most often stimulates the retina? Give an example of the production of a sight-sensation in the absence of light. What happens in the nervous system when a man "sees sparks" on receiving a blow in the eye?

the optic nerve the visual nerve-centre, and the result is a sight-sensation.*

The Eye-Socket.—The eyeball is lodged in a bony cavity, *the orbit*, open in front. Each orbit is a pyramidal chamber containing connective tissue, blood-vessels, nerves, and much fat ; the fat forms a soft cushion on which the back of the eyeball rolls.

The Eyelids are folds of skin, strengthened by cartilage and moved by muscles. Opening along the edge of each eyelid are from twenty to thirty minute glands, called the Meibomian follicles. Their secretion is sometimes abnormally abundant, and then appears as a yellowish matter along the edges of the eyelids, which often dries in the night and causes the lids to be glued together in the morning. The *eyelashes* are curved hairs, arranged in one

In what is each eyeball lodged? What does the orbit contain? What are the eyelids? The Meibomian glands? Why are the eyelids sometimes stuck together in the morning? What are the uses of the eyelashes?

---

* The fact that sight-sensations may be aroused quite independently of all light acting upon the eye is paralleled by similar phenomena in regard to other senses, and is of fundamental psychological and metaphysical importance. That a blow on the closed eye gives rise to a vivid light-sensation, even in the absence of all actual light, proves that our sensation of light is quite a different thing from light itself. The visual sensory apparatus, it is true, is so constructed and protected that of all the forces of nature, light is the one which far most frequently stimulates it. But as regards the peculiarity in the quality of the sensation which leads us to classify it as " a visual sensation," that peculiarity has nothing to do with any property of light. The visual nerve-centre when stimulated causes a sight-sensation, whether it has been excited by light, or by a blow, or by electricity. Similarly the auditory brain-centre gives us a sound-sensation when stimulated by actual external sound-waves, or by a blow on the ear, or by disease of the auditory organ. One kind of energy, *light*, excites more often than any other the visual nerve-apparatus; another, *sound*, the auditory nerve-apparatus; a third, *pressure*, the touch nerve-organs. Hence we come to associate light with visual sensations and to think of it as something like our sight-feelings; and to imagine sound as something like our auditory sensations; and so forth. As a matter of fact both light and sound are merely movements of ether or air; it is our own stimulated nerve-centres which produce visual and auditory sensations; the ethereal or aerial vibrations merely act as the stimuli which arouse the nervous apparatus.

or two rows along each lid and helping to keep dust from falling into the eye; and, when the lids are nearly closed, to protect it from a dazzling light.

The Lachrymal Apparatus consists of the tear-gland in each orbit, of ducts which carry its secretion to the upper eyelid, and of canals by which this, unless when excessive, is carried off from the front of the eye without running down over the face. The *lachrymal* or *tear gland*, about the size of an almond, lies in the upper and outer corner of the orbit. It is a compound racemose gland, from which twelve or fourteen ducts run and open at the outer corner of the upper eyelid on its inner surface. The secretion there poured out is spread evenly over the exposed part of the eye by the movements of winking, and keeps it moist; finally it is drained off by two *lachrymal canals*, one of which opens by a small pore on an elevation, or papilla, near the inner end of the margin of each eyelid. The aperture of the lower canal can be readily seen by examining its papilla in front of a looking-glass. The canals run inwards and open into the *lachrymal sac*, which lies just outside the nose, in a hollow where the lachrymal and superior maxillary bones (*L* and *Mx*, Fig. 16) meet. From this sac the *nasal duct* proceeds to open into the nose-chamber below the inferior turbinate bone (*q*, Fig. 41, p. 151).

Tears are constantly being secreted, but ordinarily in such quantity as to be drained off into the nose, from which they flow into the pharynx and are swallowed. When the lachrymal duct is stopped up, however, their

Of what parts does the lachrymal apparatus consist? Describe the lachrymal gland. Where do its ducts open? How is the front of the eyeball kept moist? Describe the arrangement by which the tears are usually carried off.

continual presence makes itself unpleasantly felt, and
may need the aid of a surgeon to clear the passage. In
weeping the secretion is increased, and then not only more
of it enters the nose, but some flows down the cheeks.
The frequent swallowing movements of a crying child,
sometimes spoken of as "gulping down his passion," are
due to the need of swallowing the extra tears which reach
the pharynx.

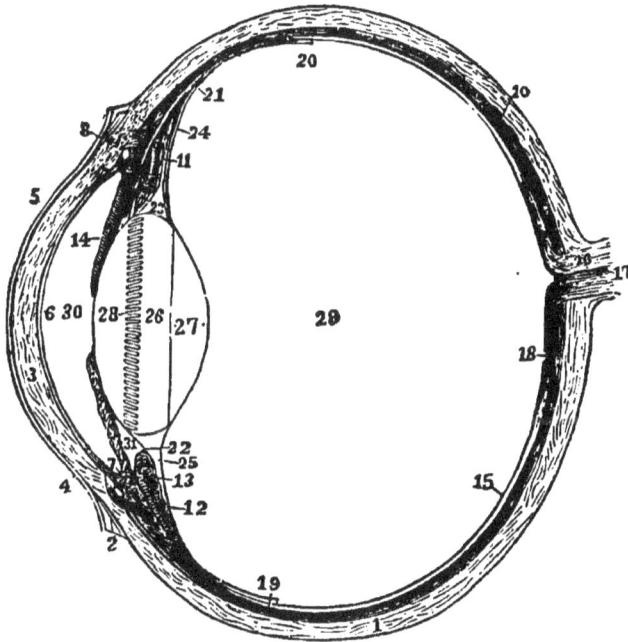

Fig. 88.—The left eyeball in horizontal section from before back. 1, sclerotic; 2, junction of sclerotic and cornea; 3, cornea; 4, 5, conjunctiva; 6, posterior elastic layer of cornea; 7, ciliary muscle; 10, choroid; 11, 13, ciliary processes; 14, iris; 15, retina; 16, optic nerve; 17, artery entering retina in optic nerve; 18, fovea centralis; 19, region where sensory part of retina ends; 22, suspensory ligament; 23 is placed in the canal of Petit, and the line from 25 points to it; 24, the anterior part of the hyaloid membrane; 26, 27, 28, are placed on the lens; 28 points to the line of attachment around it of the suspensory ligament; 29, vitreous humor; 30, anterior chamber of aqueous humor; 31, posterior chamber of aqueous humor.

Why do tears run down the face during a fit of weeping?
Why does a crying child make frequent swallowing movements?

The Globe of the Eye is on the whole spheroidal, but consists of segments of two spheres (see Fig. 88), a portion of a sphere of smaller radius forming its anterior transparent part, and being set on to the front of its posterior segment, which is part of a larger sphere. In general terms it may be described as consisting of three *coats* and three *refracting media.*

The outer coat 1 and 3, Fig. 88, consists of the *sclerotic* and the *cornea*, the latter being transparent and situated in front ; the former is opaque and white and covers the back and sides of the globe and part of the front, where it is seen between the eyelids as the *white* of the eye. Both are tough and strong, being composed of dense connective tissue.

The second coat consists of the *choroid*, 9, 10, and the *iris*, 14. The *choroid* consists mainly of blood-vessels supported by loose connective tissue, which in its inner layers contains many dark brown or black pigment granules.* Towards the front of the eyeball, where it begins to diminish in diameter, the choroid separates from the sclerotic and turns in to form the *iris*, or that colored part of the eye which is seen through the cornea ; in the centre of the iris is a circular aperture, the *pupil*, through which light reaches the interior of the eyeball.

The third or innermost coat of the eye, *the retina*, 15, is its essential portion, being the part in which the light produces those changes that give rise to nervous impulses in the optic nerve. It lines the posterior half of the eyeball.

The Microscopic Structure of the Retina is very com-

What is the form of the globe of the eye? Of what does it consist? Describe the outer coat.   The second coat.   What is the retina?

---

* In pink-eyed rabbits and in the pink-eyed ladies of "dime museums" this pigment is absent.

plex ; although but $\frac{1}{80}$ inch in thickness it presents ten distinct layers.

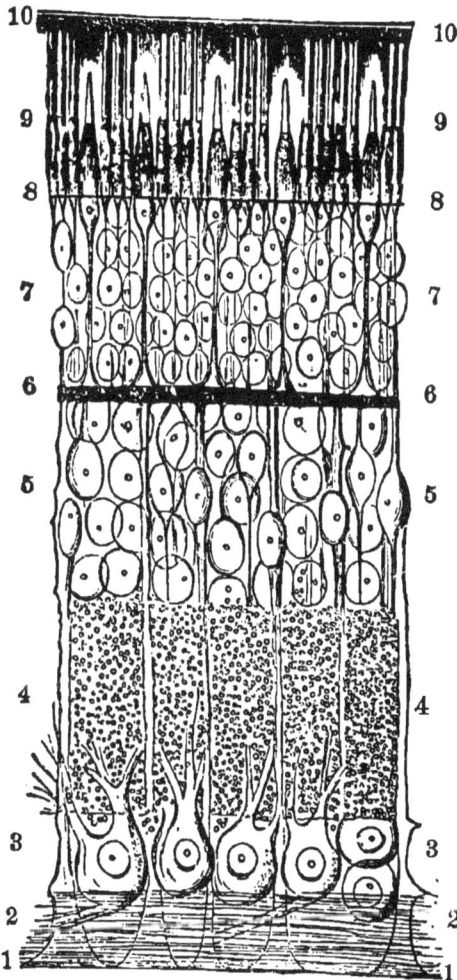

Fig. 89.—A section through the retina from its anterior or inner surface, 1 in contact with the hyaloid membrane, to its outer, 10, in contact with the choroid. 1, internal limiting membrane; 2, nerve-fibre layer; 3, nerve-cell layer; 4, inner molecular layer; 5, inner granular layer; 6, outer molecular layer; 7. outer granular layer; 8, external limiting membrane; 9, rod and cone layer; 10, pigment-cell layer.

Beginning (Fig. 89) on its front or inner side we find,

first, the *internal limiting membrane,* 1, a thin structure-less layer. Next comes the *nerve-fibre layer,* 2, formed by radiating fibres of the optic nerve; third, the *nerve-cell layer,* 3; fourth, the *inner molecular layer,* 4, consisting partly of very fine nerve-fibrils, and largely of connective tissue; fifth, the *inner granular layer,* 5, composed of nu-cleated cells, with a small amount of protoplasm at each end, and a nucleolus. These *granules,* or at any rate the majority of them, have an *inner process* running to the in-ner molecular layer, and an outer running to, 6, the *outer molecular layer,* which is thinner than the inner. Then comes, seventh, the *rod and cone fibre layer,* 7, or *outer granular layer,* composed of thick and thin fibres on each of which is a conspicuous nucleus with a nucleolus. Next is the thin *external limiting membrane,* 8, perforated by apertures through which the *rods and cones,* 9, of the ninth layer join the fibres of the seventh. Outside of all, next the choroid, is the *pigmentary layer,* 10. The nerve-fibres are believed to be continuous with the rods and cones. Light entering the eye passes through the transparent retina until it reaches the rods and cones and excites these, and they stimulate the nerves.

The action of the light is probably in the first instance chemical. The rods are stained by a purple substance, which is bleached by light and regenerated in the dark. In the healthy eye the purple is reproduced nearly as fast as de-stroyed, by the pigment-cells of the retina. Parts of the retina which contain none of this *vision purple* can see, but they may possess uncolored substances which are changed by light.

Describe the microscopic structure of the retina. On what con-stituents of the retina does light first act, in producing a sensation? With what are the rods of the retina stained? How is the pig-ment reproduced?

The Blind Spot.—Where the optic nerve enters the retina it forms a small elevation (Fig. 90), from which nerve-

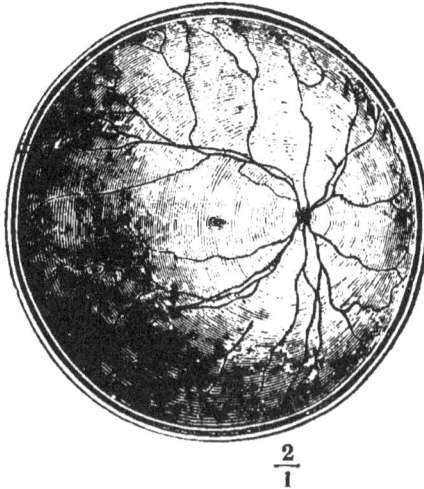

$$\frac{2}{1}$$

FIG. 90.—The right retina as it would be seen if the front part of the eyeball
with the lens and vitreous humor were removed. The white disc to the right
marks the entry of the optic nerve (blind spot); the lines radiating from this are
the retinal arteries and veins. The small central dark patch is the *yellow spot*,
the region of most acute vision.

fibres radiate. This elevation is quite blind, because it possesses neither rods nor cones. Its blindness may be readily
demonstrated. Close the left eye and look steadily with the
right at the cross (Fig. 91), holding the page vertically in

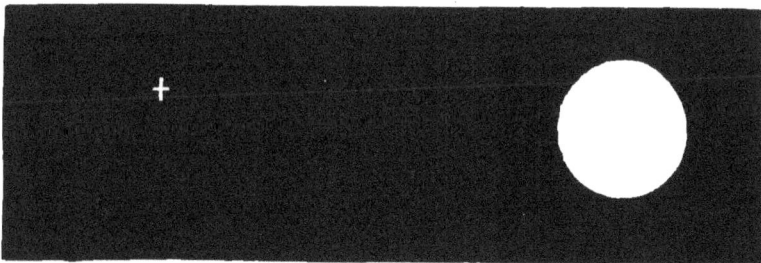

FIG. 91.

What is meant by the "blind spot"? Describe a method of
demonstrating its blindness.

front of the face, and moving it alternately from and towards
you. The eye must all the time be kept looking fixedly at
the cross. When the book is about ten inches from the
eye the white disc entirely disappears from view: its image
then falls on the part of the retina where the optic nerve
enters, and causes no visual sensation.

Light consists of vibrations in an ether which pervades
space. An object which sets up no waves in the ether does
not excite the visual nervous apparatus, and appears black;
an object which sets up ethereal vibrations capable of
exciting the rods and cones of the retina appears white
or colored when we look at it. The ethereal vibrations
enter the eye through the cornea, pass on through the
pupil, and reach and stimulate the retina.

The Refracting Media of the Eye are three in number:
(1) the *aqueous humor;* (2) the *crystalline lens;* (3) the
vitreous humor. Their relative positions are shown in Fig.
88. These media act like a convex lens, such as a commen

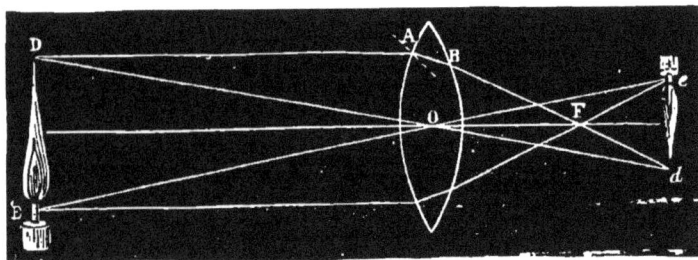

FIG. 92.—Illustrating the formation behind a convex lens of a diminished
and inverted image of an object placed in front of it.

burning-glass, and bend the rays of light which pass through
them (Fig. 92), so that all those which start from one point
of an external object meet again in a *focus* on one point of

What is light? When does an object appear black?

Name the refracting media of the eye. State their relative position. Describe their action.

the retina. In this way a small and inverted image of the things at which we look is formed on the retina, and stimulates its rods and cones.

Accommodation.—In the healthy eyeball the crystalline lens is controlled by muscles which change its convexity, making this greater when we look at near objects, and less **when** we look at distant objects. When the lens is very

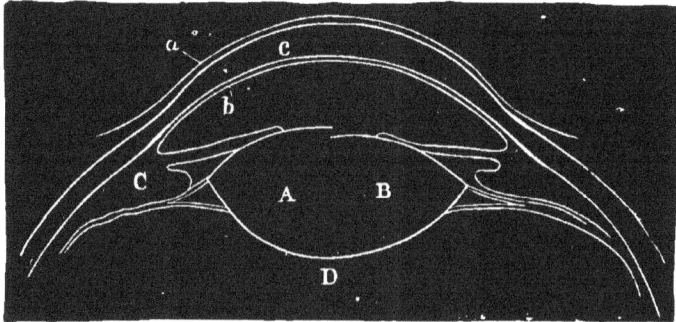

Fig. 93.—Section of front part of eyeball showing the change in the form of the lens when near and distant objects are looked at. *a, c, b,* cornea; *A,* lens when near object is looked at; *B,* lens when distant object is looked at.

convex we cannot see a distant object distinctly, and when it is less convex we only dimly see a near object. For example, standing at a window behind a lace curtain we can *look at* the curtain and see its threads plainly, but while so doing we only see indistinctly houses on the other side of the street; because the convexity of the lens is then such as to focus light from the near object on the retina, and not that from the distant. We can, however, "look at" the houses over the way and see them plainly; but then we no longer see the curtain distinctly, because the lens has so changed its form as to focus light from the far object on the

When we look at an object, what is formed on the retina?
How is the form of the crystalline lens controlled? When is its convexity greater? Can we see near and distant objects distinctly at the same moment? Illustrate.

retina, instead of light from the near. The power of changing the form of the lens according as near or distant objects
are looked at is called " *accommodation.*"

Short Sight and Long Sight.—In the normal eye the range
of accommodation is very great, allowing light from objects
infinitely distant up to
that proceeding from
those only about eight
inches in front of the eye
to be brought to a focus
on the retina. In the
natural healthy eye par-
allel rays of light meet on
the retina when the mus-
cles controlling the crystalline lens are at rest and
the lens is at its flattest
(A, Fig. 94). Such eyes
are *emmetropic.* In other
eyes the eyeball is too
long from before back;
in the resting state paral-

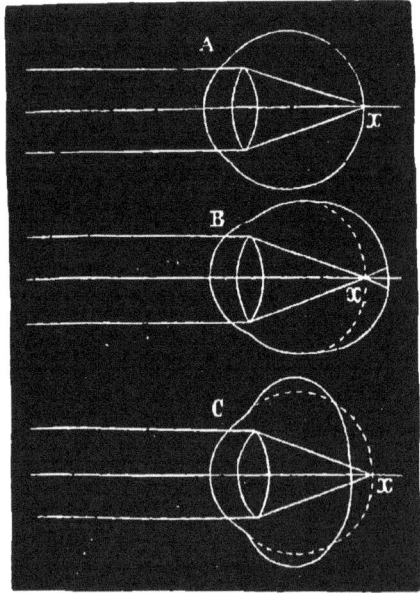

Fig. 94.—Diagram illustrating the path
of parallel rays after entering an emmetropic (A), a myopic (B), and a hypermetropic (C) eye.

lel rays meet in front of the retina (*B*). Persons with
such eyes cannot see distant objects distinctly without the
aid of diverging (concave) spectacles; they are *short-sighted*
or *myopic.* Or the eyeball may be too short from before
back; then, in its resting state, parallel rays are brought

What is meant by the "accommodation" of the eyeball?
Where do parallel rays of light which have entered a healthy eye
meet in a focus? Where do such rays meet when the eyeball is too
long from back to front? What is the result as regards vision?
What form of spectacle lenses do short-sighted persons require? Explain what is meant by a hypermetropic or long-sighted eye. What
sort of spectacles do long-sighted persons require?

to a focus behind the retina (*C*). To see even distant objects, such persons must therefore use muscular effort to increase the converging power of the lens; and when objects are near they cannot, with the greatest effort, bring the rays proceeding from them to a focus soon enough. To get distinct retinal images of near objects, they therefore need converging (convex) spectacles. Such eyes are called *hypermetropic,* or in common language *long-sighted.*

Hygiene of the Eyes.—The healthy eye is so constructed that when its muscles are at rest distinct images of distant objects are focussed on the retina. To see near objects muscular effort is required; hence the greater fatigue which follows long gazing at them.

In a hypermetropic eye more effort is needed to see near objects, and this results in muscular fatigue. Hypermetropic persons can often read well for a while, but then complain that they can no longer see distinctly. This kind of weak sight should always lead to examination of the eyes by an oculist, to see if glasses are needed; otherwise severe neuralgic pains about the eyes are apt to come on, and the overstrained organ may be permanently injured.

Children sometimes have hypermetropic eyes, and in that case should be at once provided with suitable spectacles. In old age another kind of long-sightedness (*presbyopia*) is common: it is due to too great stiffness of the crystalline lens, which does not become convex enough during accommodation to focus on the retina the images of near objects.

Short-sighted eyes appear to be more common now than formerly, especially in those given to indoor

Why is the eye apt to be fatigued by the continued contemplation of near objects?

How does the hypermetropic eye differ from the normal in the above respect? What should be done at once if a child is found to have hypermetropic eyes? What is presbyopia?

In what classes of persons is myopia most frequent?

pursuits. Myopia is rare among those who cannot read or who live mainly out of doors. It is not so apt to lead to permanent injury of the eye as hypermetropia, but the effort to see distinctly **any** but near objects is apt to produce headaches and other symptoms of nervous exhaustion. If the myopia becomes gradually worse, the eyes should be rested for several months.

Hearing.—The auditory organ (Fig. 95) consists of three portions, known respectively as the *external ear*, the *middle*

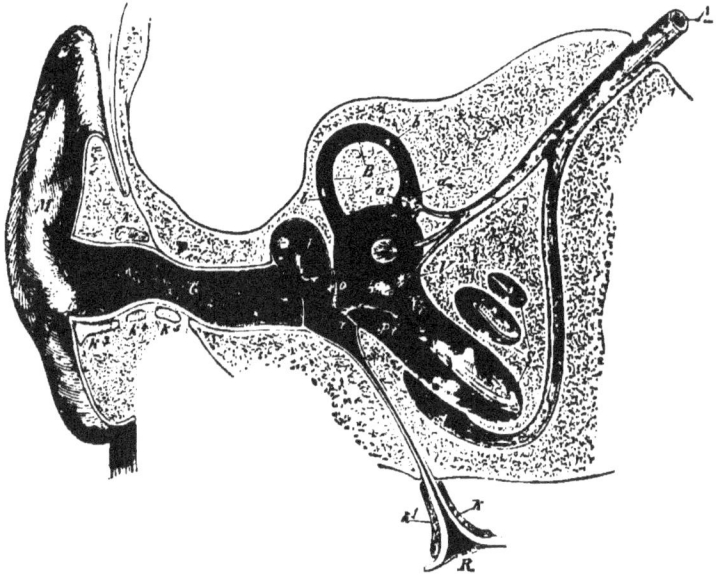

Fig. 95.—Semi-diagrammatic section through the right ear. *M*, concha. *G*, external auditory meatus. *T*, tympanic or drum membrane. *P*, Tympanum. *o*, oval foramen. *r*, round foramen. Extending from *T* to *o* is seen the chain of tympanic bones. *R*, Eustachian tube. *V, B, S*, bony labyrinth: *V*, vestibule; *B*, semicircular canal; *S*, cochlea. *b, l, l'*, membranous semicircular canal and vestibule. *A*, auditory nerve dividing into branches for vestibule, semicircular canal, and cochlea.

ear or *tympanum*, and the *internal ear* or *labyrinth;* of these the latter is the essential one, containing the ends of the auditory nerve-fibres.

What is apt to result from short sight?
Of what does the auditory organ consist?

The External Ear consists of the expansion, *M*, seen on the exterior of the head, called the *concha*, and a passage leading in from it, the *external auditory meatus*, *G*. This passage is closed at its inner end by the *tympanic* or *drum membrane*, *T*. It is lined by a prolongation of the skin, through which numerous small glands, secreting the *wax* of the ear, open.

The Tympanum, or drum chamber of the ear (Fig. 96 and *P*, Fig. 95), is an irregular cavity in the temporal bone, closed externally by the drum membrane. From its inner side the *Eustachian tube* (*R*, Fig. 95) proceeds and opens into the pharynx. This tube allows air from the throat to enter the tympanum, and serves to keep equal the atmospheric pressure on each side of the drum

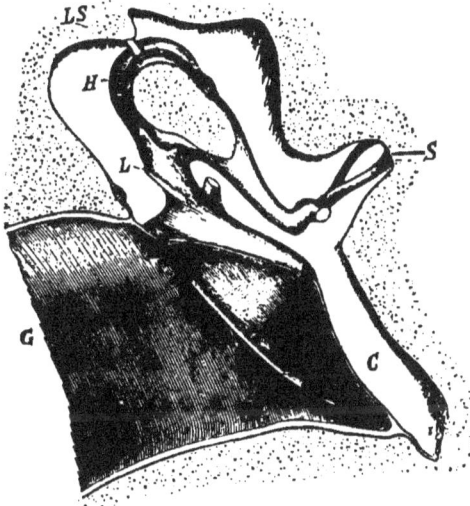

Fig. 96.—The tympanic cavity and its bones, considerably magnified. *G*, the inner end of the external auditory meatus, closed internally by the conical tympanic membrane; *L*, the malleus, or hammer-bone; *H*, the incus, or anvil-bone; *S*, the stapes, or stirrup-bone.

membrane. Three small bones (Fig. 96) stretch across the tympanic cavity from the drum membrane to the labyrinth; they transmit the vibrations of the membrane, produced by sound-waves in the air, to the liquid of the

Describe the external ear.

What is the tympanum? The Eustachian tube? What is the use of the Eustachian tube? What bones lie in the tympanum? What is their function?

labyrinth. The outmost bone is the *malleus ;* the inmost, the *stapes ;* and the middle bone, the *incus.*

The Internal Ear, or Labyrinth, consists primarily of chambers and tubes hollowed out in the temporal bone. The middle chamber, called the *vestibule* (*V,* Fig. 95), has an opening, the *oval foramen, o,* in its outer side, into which the inner end of the stapes, or stirrup-bone, fits. Behind, the vestibule opens into three *semicircular canals,* one of which is shown at *B,* Fig. 95; and in front into a spirally coiled tube, *S,* the *cochlea.* In these bony chambers and tubes lie membranous chambers and tubes, in which the fibres of the auditory nerve (*A,* Fig. 95) end. All the labyrinth chamber outside these membranous parts is occupied by a watery liquid, known as *perilymph;* the membranous chambers are filled with a similar liquid, the *endolymph.*

When sound-waves of the air make the tympanic membrane vibrate, it shakes the tympanic bones; the stapes then shakes the liquids in the labyrinth, and sets up vibrations in them, which excite the endings of the auditory nerve. The stimulated auditory nerve then conveys a nervous impulse to the brain-centre of hearing and excites it, and a sensation of sound results.

Touch, or the Pressure Sense.—Many sensory nerves end in the skin, and through it we get several kinds of sensa-

Of what does the internal ear primarily consist? What is the vestibule?
What is found on the outer side of the vestibule?
Into what does the vestibule open behind? In front? What lie in the bony cavities of the labyrinth? What is the perilymph? The endolymph?
What happens when sound-waves set the tympanic membrane in vibration?

tion; *touch* proper, *heat* and *cold*, and *pain ;* and we can with more or less accuracy localize them on the surface of the body. The interior of the mouth possesses also these sensibilities. Through touch proper we recognize pressure or traction exerted on the skin, and the force of the pressure; the softness or hardness, roughness or smoothness, of the body producing it; and the form of this, when not too large to be felt all over.

The delicacy of the tactile sense varies on different parts of the skin; it is greatest on the forehead and temples, where a weight of $\frac{3}{100}$ of a grain can be felt.

The Localization of Skin Sensations.—When the eyes are closed and a point of the skin is touched we can with some accuracy indicate the region stimulated; although tactile feelings are alike in general characters, they differ in something (*local sign*) besides intensity by which we can distinguish them as originated on different parts of the skin. The accuracy of the localizing power varies widely in different skin regions, and is measured by observing the least distance which must separate two objects (as the blunted points of a pair of compasses) in order that they may be felt as two. The following table illustrates some of the differences observed:

Tongue-tip	.04 inch
Palm side of last phalanx of finger	.08 "
Red part of lips	.16 "
Tip of nose	.24 "
Back of second phalanx of finger	.44 "
Heel	.88 "

What sensations do we get through the skin? Name another part of the body which also gives rise to these sensations. What do we recognize by means of the sense of touch?

Where is the tactile sense most acute?

What is meant by the localization of skin sensations? Does the accuracy of localization differ in different regions of the skin? Illustrate.

Back of hand....................................	1.23 inches
Forearm...	1.58 "
Sternum..	1.76 "
Back of neck....................................	2.11 "
Middle of back..................................	2.64 "

The Temperature Senses.—By these is meant our faculty of perceiving cold and warmth; and, with the help of these sensations, of perceiving temperature differences in external objects. The organs are the skin, the mucous membrane of mouth, pharynx, and gullet, and of the entry of the nose. Burning the skin will cause pain, but not a true temperature sensation, which is quite as different from pain as is touch.

Smell.—The olfactory organ consists of the mucous membrane of the upper parts of the nasal cavities; in it the endings of the olfactory nerves are spread. It covers the upper and lower turbinate bones (*o, p,* Fig. 41) (which are expansions of the ethmoid on the outer wall of the nostril-chamber), the opposite part of the partition between the nares, and that part of the roof of the nose (*n,* Fig. 41) which separates it from the cranial cavity.

Odorous Substances, the stimuli of the olfactory apparatus, are always gaseous. They frequently act powerfully when present in very small quantity. A grain or two of musk kept in a room will give the air in it an odor for years, and yet at the end will hardly have diminished in weight, so infinitesimal is the quantity given off from it to the air and able to excite the sense of smell. While some gases or

What is the temperature sense? What are its organs? Of what does the olfactory organ consist?
In what point do all odorous substances agree? Illustrate the efficiency, so far as producing smell sensations is concerned, of a very small quantity of an odorous substance. Do all gases stimulate the olfactory apparatus?

vapors have this powerful influence upon the olfactory organ, others, as pure air, do not stimulate it at all.

Taste.—The organ of taste is the mucous membrane on the upper side of the tongue, and possibly on other parts of the boundary of the mouth-cavity. The mucous membrane of the tongue presents innumerable elevations or papillæ (Fig. 46) of three kinds (p. 159). The filiform papillæ are organs of touch, for the tongue has the sense of touch as well as of taste. The circumvallate and fungiform papillæ contain the endings of branches of the glosso-pharyngeal and trigeminal nerves (pp. 325, 324), which, when excited by sapid bodies, stimulate the taste-centres in the brain.

Many so-called tastes (flavors) are really smells; odoriferous particles of substances which are being eaten reach the nose through the posterior nares and arouse smell sensations which, since they accompany the presence of objects in the mouth, we take for tastes. Such is the case with most spices; when the nasal chambers are blocked or inflamed by a cold in the head, or closed by pinching the nose, the so-called "taste" of spices is not perceived when they are eaten; all that is felt when cinnamon, *e.g.*, is chewed under such circumstances is a certain pungency due to its stimulation of nerves of common sensation in the tongue. This fact is sometimes taken advantage of in the practice of domestic medicine when a nauseous dose, as rhubarb, is to be given to a child.

What is the organ of taste? What is found on the mucous membrane of the tongue? What papillæ are concerned in the tactile sensibility of the tongue? In the gustatory? What nerves supply the taste-papillæ?

What are many so-called tastes? Illustrate.

CHAPTER XXIV.

VOICE AND SPEECH.

Voice consists of sounds produced by the vibrations of two elastic bands called the *vocal cords*. These cords lie in the *larynx*, which is situated between the pharynx and the windpipe, and is a portion of the passage conveying air to the lungs specially modified to form a voice-organ.

The vocal cords project into the larynx so that but a narrow slit, called the *glottis*, is left between them. When the vocal cords are put in a certain position air driven through the glottis sets them vibrating and they give origin to sounds. The stronger the blast the louder the voice.

The *pitch* of the voice is primarily dependent on the size of the larynx. The larger it is, or what comes to the same thing, the longer the vocal cords are, the lower is the pitch of the voice. In children, therefore, the voice is shrill; and, as the female larynx is usually smaller than the male, a woman's voice is usually higher pitched than a man's. About sixteen or seventeen years of age a boy's larynx grows very fast, and his voice "breaks," becoming about an octave deeper in tone.

How is voice produced? Where do the vocal cords lie? Where is the larynx situated?

What is the glottis? When do the vocal cords give origin to sounds? On what does the loudness of the voice depend?

How does the size of larynx influence the pitch of the voice? Why is a woman's voice commonly higher pitched than a man's? Why does a youth's voice break?

While every one's voice has a certain natural pitch which leads us to call it soprano, tenor, bass, and so forth, this pitch can be modified within limits, so that we each can sing a number of notes. This variety is due to the action of muscles in the larynx which alter the tension of the vocal cords; the more tightly these are stretched, other things being equal, the higher pitched is the tone which they emit.

Speech.—The vocal cords alone would produce but feeble sounds. If a fiddle-string be attached to a hook on the ceiling and stretched by hanging a heavy weight on its lower end, we can get tones out of it when it is plucked or bowed; but the tones are feeble and deficient in character and fullness. In the violin the strings are attached to a hollow wooden box, and when the string is set in movement it causes the wood to vibrate, and this, in turn, the air contained in the cavity of the instrument; in this way the tone is intensified, and altered and much improved in quality. The air in the pharynx, mouth, and nose answers pretty much to that in the hollow of the violin; those cavities together form a resonance-chamber, and when the vocal cords vibrate they set this air in vibration also, and so the sound is made louder and is altered in character. By movements of throat, soft palate, tongue, cheeks, and lips, the size and form of the sounding chamber are varied, and with them the tone of voice; by movements of tongue, lips, and palate, the air-current, and therefore the sound, is interrupted from time to time; on other occasions the

How is it that we can sing a number of notes of different pitch? Why is a hollow wooden box an essential part of a violin? How do the throat and mouth cavities influence the loudness and quality of the voice? How do tongue, lips, and cheeks co-operate in converting voice into speech?

air is forced through a narrow passage in the mouth, giving
rise to new sounds added on to those originated by the vocal
cords. In such manners the primitive feeble monotonous
tone due to the vocal cords is reinforced and altered in
various ways in throat and mouth, and *voice* is developed
into articulate *speech*.

The **Larynx** consists of a framework of nine cartilages,

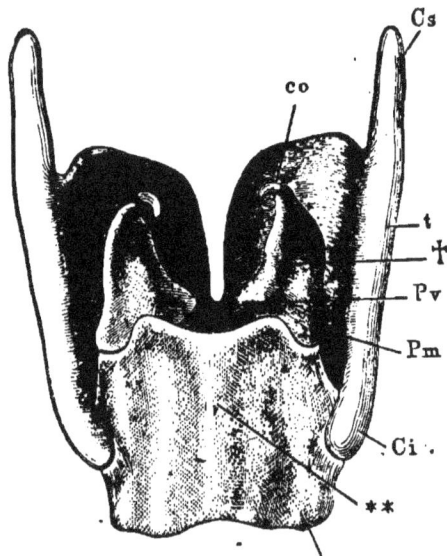

FIG. 97.—The more important cartilages of the larynx from behind. *t*, thyroid; *Cs*, its superior, and *Ci*, its inferior, horn of the right side; **, cricoid cartilage; †, arytenoid cartilage; *Pv*, the corner to which the posterior end of a vocal cord is attached; *Pm*, corner on which the muscles which approximate or separate the vocal cords are inserted; *co*, cartilage of Santorini.

movably articulated together, and having muscles attached
to them by whose contractions their relative positions are
altered; the cartilages surround a tube, continuous below
with the windpipe, and lined by mucous membrane. At
one level in the laryngeal tube the vocal cords project and,

Of what does the laryngeal framework consist? What do the
cartilages of the larynx surround? How is the glottis formed?

pushing out the mucous membrane, leave for the passage of air only the narrow slit of the glottis, above mentioned.

The largest cartilage of the larynx (*t*, Fig. 97) is the *thyroid*. It is placed in front and consists of right and left halves which meet at an angle in front, but separate behind so as to enclose a V-shaped space. The front of the thyroid cartilage causes the prominence in the neck known as Adam's apple. The *epiglottis*, not represented in the figure, is attached to the top of the thyroid cartilage and overhangs the entry from pharynx to larynx. It may be seen, covered by mucous membrane, projecting at the root of the tongue, if that organ be pushed down, while the mouth is held open before a mirror. It is represented as seen from behind at *a*, Fig. 98. The *cricoid cartilage* (★★, Fig. 97) has the form of a signet-ring, with its broad part turned towards the back of the throat, and placed in the lower part of the opening between the halves of the thyroid. The two *arytenoid cartilages* (†, Fig. 97) are placed on the top of the wide posterior part of the cricoid; each is pyramidal in form. The remaining laryngeal cartilages are of less importance.

The Vocal Cords, which are rather projecting pads of elastic tissue than cords in the ordinary sense of the word, proceed, one from each arytenoid cartilage behind, to the angle where the halves of the thyroid meet in front. In quiet breathing the interval (*glottis*) between them (*c*,

Describe the position and form of the thyroid cartilage. What causes "Adam's apple"? What is the epiglottis? How may you see it in your own throat? Describe the cricoid cartilage. What cartilages are set on top of the cricoid? What is their form?

Between what points are the vocal cords stretched? Under what circumstances does air driven through the glottis not set them vibrating?

Fig. 98) is narrow in front and wider behind: under such circumstances air driven through the opening does not set the margins of the cords in vibration, and no sound is produced.

The Muscles of the Larynx. The laryngeal muscles

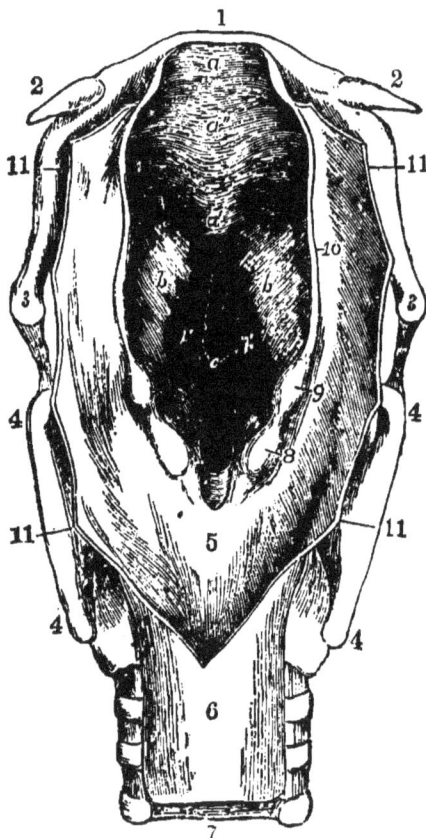

FIG. 98.—The larynx viewed from its pharyngeal opening. The back wall of the pharynx has been divided and its edges (11) turned aside. 1, body of hyoid; 2, its small, and 3, its great, horns; 4, upper and lower horns of thyroid cartilage; 5, mucous membrane of front of pharynx, covering the back of the cricoid cartilage; 6, upper end of gullet; 7, windpipe, lying in front of the gullet; 8, eminence caused by cartilage of Santorini; 9, eminence caused by cartilage of Wrisberg; both lie in, 10, the *aryteno-epiglottidean fold* of mucous membrane, surrounding the opening (*aditus laryngis*) from pharynx to larynx. *a*, projecting tip of epiglottis; *c*, the glottis, the lines leading from the letter-point to the free vibrating edges of the vocal cords. *b′*, the ventricles of the larynx: their upper edges, marking them off from the eminences *b*, are the false vocal cords.

are numerous, and are arranged—(1) to pull the arytenoid cartilages towards one another and so narrow the glottis behind; then air forced through the narrowed slit sets the cords vibrating and produces voice. (2) To increase the distance between the arytenoid cartilages behind and the thyroid in front: as the vocal cords are attached to both, this action stretches and tightens them, and so raises the pitch of the voice. (3) To pull the front of the thyroid cartilage nearer the arytenoids and so slacken the cords and lower the pitch of the voice. (4) To separate the arytenoid cartilages, and with them the vocal cords, and thus widen the glottis and allow air to pass through it without producing voice.

The Range of the Human Voice from the lowest note (*f* of the unaccented octave) of an ordinary bass to the highest note (*g* on the thrice-accented octave) of a fairly good soprano is about three octaves: the former note is produced by 88 vibrations per second, the latter by 792. Celebrated singers of course go beyond such limit in each direction: bassos have been known to take *a* on the great octave (55 vibrations per second), and Mozart, at Parma, heard a soprano sing a note of the extraordinarily high pitch *c* on the fifth accented octave (2112 vibrations per second).

Vowels are musical tones produced in the larynx and modified by resonance of the air in the pharynx and mouth. To get the broad *a* sounds, as *ah*, the mouth is widely opened and the lips drawn back; to get such vowels as

State the uses of the muscles of the larynx.
What is the ordinary range of the human voice? What notes have celebrated singers taken beyond the ordinary highest and lowest limits?
What are vowels? Illustrate the influence of the shape given to the mouth-cavity in the production of different vowels.

oo (*moor*) the lips are protruded and the mouth cavity lengthened. The change in the form of the mouth may be noticed by pronouncing consecutively the vowel-sounds *ah, eh, ee, oh, oo.* The English *i* (as in *spire*) is a diphthong, consisting of *ă* (p*a*d) followed by *ē* (f*ee*t), as may be readily found on attempting to sing a sustained note to the sound *ī.*

Semivowels.—In uttering true vowel-sounds the soft palate is raised so as to cut off the air in the nose, which then does not take part in the resonance. For some other sounds (the *semivowels* or *resonants*) the initial step is, as in the case of the true vowels, the production of a laryngeal tone; but the soft palate is not raised, and the mouth-exit is more or less closed by the lips or the tongue; hence the blast partly issues through the nose, and the air there takes part in the vibrations and gives them a special character; this is the case with *m, n,* and *ng.*

Consonants are sounds produced not mainly by the vocal cords, but by modifications of the expiratory blast on its way through the mouth. The current may be interrupted and the sound changed by the lips (*labials,* as *p* and *b*); or, at or near the teeth, by the tip of the tongue (*dentals,* as *t* and *d*); or, in the throat, by the root of the tongue and the soft palate (*gutturals,* as *k* and *g*).

Consonants may also be classified by the kind of movement which gives rise to them. In *explosives* an interruption to the air-current is suddenly interposed or removed (*p, b, t, d, k, g*). Other consonants are *continuous* (*f, s, r*) and may be divided into (1) *aspirates,* when the air

Is the long *i* of English a true vowel ?
What is meant by the semivowels ?
What are consonants ? How may they be classified ?

is made to rush through a narrow aperture, as, for example, between the lips (*f*) or the teeth (*s*) or the tongue
and the palate (*sh*) or the tongue and the teeth (*th*);
(2) *resonants* or semivowels; (3) *vibratories*, the different
forms of *r*, due to vibrations of parts bounding a constriction put in the way of the air-current on its passage.

INDEX.

THE AMERICAN SCIENCE SERIES.

THE principal objects of the series are to supply the lack—in some subjects very great—of authoritative books whose principles are, so far as practicable, illustrated by familiar American facts, and also to supply the other lack that the advance of Science perennially creates, of text-books which at least do not contradict the latest generalizations. The scheme systematically outlines the field of Science, as the term is usually employed with reference to general education, and includes ADVANCED COURSES for maturer college students, BRIEFER COURSES for beginners in school or college, and ELEMENTARY COURSES for the youngest classes. The Briefer Courses are not mere abridgments of the larger works, but, with perhaps a single exception, are much less technical in style and more elementary in method. While somewhat narrower in range of topics, they give equal emphasis to controlling principles. The following books in this series are already published ·

THE HUMAN BODY. By H. NEWELL MARTIN, Professor in the Johns Hopkins University.

Advanced Course. 8vo. 655 pp.

Designed to impart the kind and amount of knowledge every educated person should possess of the structure and activities and the conditions of healthy working of the human body. While intelligible to the general reader, it is accurate and sufficiently minute in details to meet the requirements of students who are not making human anatomy and physiology subjects of special advanced study. *The regular editions of the book contain an appendix on Reproduction and Development. Copies without this will be sent when specially ordered.*

From the CHICAGO TRIBUNE: " The reader who follows him through to the end of the book will be better informed on the subject of modern physiology in its general features than most of the medical practitioners who rest on the knowledge gained in comparatively antiquated text-books, and will, if possessed of average good judgment and powers of discrimination, not be in any way confused by statements of dubious questions or conflicting views."

THE HUMAN BODY.—*Continued.*

Briefer Course. 12mo. 364 pp.

Aims to make the study of this branch of Natural Science a source of discipline to the observing and reasoning faculties, and not merely to present a set of facts, useful to know, which the pupil is to learn by heart, like the multiplication-table. With this in view, the author attempts to exhibit, so far as is practicable in an elementary treatise, the ascertained facts of Physiology as illustrations of, or deductions from, the two cardinal principles by which it, as a department of modern science, is controlled,—namely, the doctrine of the "Conservation of Energy" and that of the "Physiological Division of Labor." To the same end he also gives simple, practical directions to assist the teacher in demonstrating to the class the fundamental facts of the science. *The book includes a chapter on the action upon the body of stimulants and narcotics.*

From HENRY SEWALL, *Professor of Physiology, University of Michigan:* "The number of poor books meant to serve the purpose of text-books of physiology for schools is so great that it is well to define clearly the needs of such a work: 1. That it shall contain accurate statements of fact. 2. That its facts shall not be too numerous, but chosen so that the important truths are recognized in their true relations. 3. That the language shall be so lucid as to give no excuse for misunderstanding. 4. That the value of the study as a discipline to the reasoning faculties shall be continually kept in view. I know of no elementary text-book which is the superior, if the equal, of Prof. Martin's, as judged by these conditions."

Elementary Course. 12mo. 261 pp.

A very earnest attempt to present the subject so that children may easily understand it, and, whenever possible, to start with familiar facts and gradually to lead up to less obvious ones. *The action on the body of stimulants and narcotics is fully treated.*

From W. S. PERRY, *Superintendent of Schools, Ann Arbor, Mich.:* "I find in it the same accuracy of statement and scholarly strength that characterize both the larger editions. The large relative space given to hygiene is fully in accord with the latest educational opinion and practice; while the amount of anatomy and physiology comprised in the compact treatment of these divisions is quite enough for the most practical knowledge of the subject. The handling of alcohol and narcotics is, in my opinion, especially good. The most admirable feature of the book is its fine adaptation to the capacity of younger pupils. The diction is simple and pure, the style clear and direct, and the manner of presentation bright and attractive."

ASTRONOMY. By SIMON NEWCOMB, Professor in the Johns Hopkins University, and EDWARD S. HOLDEN, Director of the Lick Observatory.

Advanced Course. 8vo. 512 pp.

To facilitate its use by students of different grades, the subject-matter is divided into two classes, distinguished by the size of the type. The portions in large type form a complete course for the use of those who desire only such a general knowledge of the subject as can be acquired without the application of advanced mathematics. The portions in small type comprise additions for the use of those students who either desire a more detailed and precise knowledge of the subject, or who intend to make astronomy a special study.

From C. A. YOUNG, *Professor in Princeton College :* " I conclude that it is decidedly superior to anything else in the market on the same subject and designed for the same purpose."

Briefer Course. 12mo. 352 pp.

Aims to furnish a tolerably complete outline of the astronomy of to-day, in as elementary a shape as will yield satisfactory returns for the learner's time and labor. It has been abridged from the larger work, not by compressing the same matter into less space, but by omitting the details of practical astronomy, thus giving to the descriptive portions a greater relative prominence.

From THE CRITIC: "The book is in refreshing contrast to the productions of the professional schoolbook-makers, who, having only a superficial knowledge of the matter in hand, gather their material, without sense or discrimination, from all sorts of authorities, and present as the result an *indigesta moles*, a mass of crudities, not unmixed with errors. The student of this book may feel secure as to the correctness of whatever he finds in it. Facts appear as facts, and theories and speculations stand for what they are, and are worth."

From W. B. GRAVES, *Master Scientific Department of Phillips Academy :* " I have used the Briefer Course of Astronomy during the past year. It is up to the times, the points are put in a way to interest the student, and the size of the book makes it easy to go over the subject in the time allotted by our schedule."

From HENRY LEFAVOUR. *late Teacher of Astronomy, Williston Seminary :* " The impression which I formed upon first examination, that it was in very many respects the best elementary text-book on the subject, has been confirmed by my experience with it in the class-room."

ZOOLOGY. By A. S. PACKARD, Professor in Brown University.

Advanced Course. 8vo. 719 pp.

Designed to be used either in the recitation-room or in the laboratory. It will serve as a guide to the student who, with a desire to get at first-hand a general knowledge of the structure of leading types of life, examines living animals, watches their movements and habits, and finally dissects them. He is presented first with the facts, and led to a thorough knowledge of a few typical forms, then taught to compare these with others, and finally led to the principles or inductions growing out of the facts

From A. E. VERRILL, *Professor of Zoology in Yale College:* "The general treatment of the subject is good, and the descriptions of structure and the definitions of groups are, for the most part, clear, concise, and not so much overburdened by technical terms as in several other manuals of structural zoology now in use."

Briefer Course. 12mo. 334 pp.

The distinctive characteristic of this book is its use of the *object method.* The author would have the pupils first examine and roughly dissect a fish, in order to attain some notion of vertebrate structure as a basis of comparison. Beginning then with the lowest forms, he leads the pupil through the whole animal kingdom until man is reached. As each of its great divisions comes under observation, he gives detailed instructions for dissecting some one animal as a type of the class, and bases the study of other forms on the knowledge thus obtained.

From HERBERT OSBORN, *Professor of Zoology, Iowa Agricultural College:* "I can gladly recommend it to any one desiring a work of such character. While I strongly insist that students should study animals from the animals themselves,—a point strongly urged by Prof. Packard in his preface,—I also recognize the necessity of a reliable text-book as a guide. As such a guide, and covering the ground it does, I know of nothing better than Packard's."

First Lessons in Zoology. 12mo. 290 pp.

In method this book differs considerably from those mentioned above. Since it is meant for young beginners, it describes but few types, mostly those of the higher orders, and discusses their relations to one another and to their surroundings. The aim, however, is the same with that of the others; namely, to make clear the general principles of the science, rather than to fill the pupil's mind with a mass of what may appear to him unrelated facts.

PSYCHOLOGY—Advanced Course. BY WILLIAM JAMES, Professor in Harvard University. 2 vols. 8vo., 689, 704 pp.

From Prof. E. H. GRIFFIN, *John Hopkins University:* "An important contribution to psychological science, discussing its present aspects and problems with admirable breadth, insight, and independence."

From Prof. JOHN DEWEY, *University of Michigan:* " A remarkable union of wide learning, originality of treatment, and, above all, of never-failing suggestions. To me the best treatment of the whole matter of advanced psychology in existence. It does more to put psychology in scientific position both as to the statement of established results and a stimulating to further problems and their treatment, than any other book of which I know."

From Hon. W. T. HARRIS, *National Bureau of Education:* " I have never seen before a work that brings together so fully all of the labors, experimental and analytic, of the school of physiological psychologists."

BOTANY. By CHARLES E. BESSEY, Professor in the University of Nebraska.

Advanced Course. 8vo. 611 pp.

Aims to lead the student to obtain at first-hand his knowledge of the anatomy and physiology of plants. Accordingly, the presentation of matter is such as to fit the book for constant use in the labaratory, the text supplying the outline sketch which the student is to fill in by the aid of scalpel and microscope.

From J. C. Arthur, Editor of *The Botanical Gazette:* " The first botanical text-book issued in America which treats the most important departments of the science with anything like due consideration. This is especially true in reference to the physiology and histology of plants, and also to special morphology. Structural Botany and classification have up to the present time monopolized the field, greatly retarding the diffusion of a more complete knowledge of the science."

Essentials of Botany. 12mo. 292 pp.

A guide to beginners. Its principles are, that the true aim of botanical study is not so much to seek the family and proper names of specimens as to ascertain the laws of plant structure and plant life; that this can be done only by examining and dissecting the plants themselves ; and that it is best to confine the attention to a few leading types, and to take up first the simpler and more easily understood forms, and afterwards those whose structure and functions are more complex.

From J. T. ROTHROCK, *Professor in the University of Pennsylvania:* " There is nothing superficial in it, nothing needless introduced, nothing essential left out. The language is lucid ; and, as the crowning merit of the book, the author has introduced throughout the volume ' Practical Studies,' which direct the student in his effort to see for himself all that the text-book teaches."

CHEMISTRY. By IRA REMSEN, Professor in the Johns Hop kins University.

Advanced Course. 8vo.   828 pp.

The general plan of this work will be the same with that of the Briefer Course, already published. But the part in which the members of the different families are treated will be con siderably enlarged. Some attention will be given to the lines of investigation regarding chemical affinity, dissociation, speed of chemical action, mass action, chemical equilibrium, thermo-chemistry, etc. The periodic law, and the numerous relations which have been traced between the chemical and physical properties of the elements and their positions in the periodic system will be specially emphasized. Reference will also be made to the subject of the chemical constitution of compounds, and the methods used in determining constitution.

Introduction to the Study of Chemistry. 12mo.   389 pp.

The one comprehensive truth which the author aims to make clear to the student is the essential nature of chemical action. With this in view, he devotes the first 208 pages of the book to a carefully selected and arranged series of simple experiments. in which are gradually developed the main principles of the subject. His method is purely inductive ; and, wherever experience has shown it to be practicable, the truths are drawn out by pointed questions, rather than fully stated. Next, when the student is in a position to appreciate it, comes a simple account of the theory of the science. The last 150 pages of the book are given to a survey, fully illustrated by experiments, of the leading families of *inorganic* compounds.

From ARTHUR W. WRIGHT, *Professor in Yale College :*—The student is not merely made acquainted with the phenomena of chemistry, bu is constantly led to reason upon them, to draw conclusions from them and to study their significance with reference to the processes o. chemical action—a course which makes the book in a high degree dis-ciplinary as well as instructive.

From THOS. C. VAN NUYS, *Professor of Chemistry in the Indiana University :*—It seems to me that Remsen's "Introduction to the Study of Chemistry" meets every requirement as a text or class book.

From C. LES MEES, *Professor of Chemistry in the Ohio University.* —I unhesitatingly recommend it as the best work as yet published for the use of beginners in the study. Having used it. I feel justified in saving this much.

CHEMISTRY—*Continued.*

Elements of Chemistry. 12mo. 272 pp.

Utilizes the facts of every-day experience to show what chemistry is and how things are studied chemically. The language is untechnical, and the subject is fully illustrated by simple experiments, in which the pupil is led by questions to make his own inferences. The author has written under the belief that "a rational course in chemistry, whether for younger or older pupils, is something more than a lot of statements of facts of more or less importance; a lot of experiments of more or less beauty; or a lot of rules devised for the purpose of enabling the pupil to tell what things are made of. If the course does not to some extent help the pupil to think as well as to see it does not deserve to be called rational."

CHASE PALMER, *Professor in the State Normal School, Salem, Mass.:* —It is the best introduction to chemistry that I know, and I intend to put it into the hands of my pupils next Fall.

A. D. GRAY, *Instructor in Springfield (Mass.) High School :*—Neat, attractive, clear, and accurate, it leaves little to be desired or sought for by one who would find the best book for an elementary course in our High Schools and Academies.

GENERAL BIOLOGY. By WILLIAM T. SEDGWICK, Professor in the Mass. Institute of Technology, and EDMUND B. WILSON, Professor in Bryn Mawr College. *Part I.* 8vo. 193 pp.

This work is intended for college and university students as an introduction to the theoretical and practical study of biology. It is not zoology, botany, or physiology, and is intended not as a substitute, but as a foundation, for these more special studies. In accordance with the present obvious tendency of the best elementary biological teaching, it discusses broadly some of the leading principles of the science on the substantial basis of a thorough examination of a limited number of typical forms, including both plants and animals. Part First, now published, is a general introduction to the subject illustrated by the study of a few types. Part Second will contain a detailed survey of various plants and animals.

W. G. FARLOW, *Professor in Harvard University, Cambridge, Mass.:* —An introduction is always difficult to write, and I know no work in which the general relations of plants and animals and the cell-structure have been so well stated in a condensed form.

POLITICAL ECONOMY. By Francis A. Walker, President of the Massachusetts Institute of Technology.

Advanced Course. 8vo. 537 pp.

The peculiar merit of this book is its *reality*. The reader is brought to see the application of the laws of political economy to real facts. He learns the extent to which those laws hold good, and the manner in which they are applied. The subject is divided, as usual, into the three great branches of production, exchange, and distribution. An interesting and suggestive "book" on consumption is added, which serves to bring in conveniently the principles of population. The last part of the volume is given to the consideration of various practical applications of economic principles.

From Richmond Mayo Smith, *Professor in Columbia College, N. Y.:*—In my opinion it is the best text-book of political economy that we as yet possess.

From Woodrow Wilson, *Professor in Princeton University, N. J.:* —It serves better than any other book I know of as an introduction to the most modern point of view as to economical questions.

Briefer Course. 12mo. 415 pp.

The demand for a briefer manual by the same author for the use of schools in which only a short time can be given to the subject has led to the publication of the present volume. The work of abridgment has been effected mainly through excision, although some structural changes have been made, notably in the parts relating to distribution and consumption.

From Alexander Johnston, *late Professor in Princeton University, N. J.:*—Using the "Briefer Course" as a text-book, suited to any capacity, I am able at the same time to recommend the "Advanced Course" to those who are better able to use it as a book of reference, or more inclined to carry their work further.

Elementary Course. 12mo. 323 pp.

What has been attempted is a clear arrangement of topics; a simple, direct, and forcible presentation of the questions raised; the avoidance, as far as possible, of certain metaphysical distinctions which the author has found perplexing; a frequent repetition of cardinal doctrines, and especially a liberal use of concrete illustrations, drawn from facts of common experience or observation.

HENRY HOLT & CO PUBLISHERS, N. Y.